Manfred Walk

Theory of Duality in Mathematical Programming

Springer-Verlag Wien New York

Prof. Dr. Manfred Walk
Friedrich-Schiller-Universität
Sektion Mathematik
Jena, German Democratic Republic

This work is published simultaneously
by Akademie-Verlag Berlin
and by Springer-Verlag Wien—New York
and is subject to copyright.

Distribution rights for Socialist Countries:
Akademie-Verlag Berlin

Distribution rights for all countries excluding Socialist Countries:
Springer-Verlag Wien—New York

With 9 Figures

Library of Congress Cataloging-in-Publication Data. Walk, M. (Manfred), Theory of duality in mathematical programming. Bibliography: p. Includes index. 1. Programming (Mathematics). 2. Duality theory (Mathematics). I. Title. QA402.5.W33. 1989. 519.7. 88-6575.

Printed in the German Democratic Republic
Printing and binding by:
VEB Druckhaus „Maxim Gorki", DDR-7400 Altenburg

ISBN 3-211-82057-4 Springer-Verlag Wien—New York
ISBN 0-387-82057-4 Springer-Verlag New York—Wien
ISBN 3-05-500404-3 Akademie-Verlag Berlin

Preface

The theory of mathematical programming has developed considerably during the last years due to the stimulus given by an increasing demand from management and technology and it is therefore understandable that courses on mathematical programming now belong to standard teaching programs of Universities and Institutes.

Our aim in writing this book is to introduce students of mathematics, science, economics and management to the qualitative theory of mathematical programming in vector spaces. Prerequisite for the study of this book is a basic knowledge of analysis and linear algebra. We also apply some elementary ideas of functional analysis for a more rigorous construction of proofs and for some generalizations of the finite dimensional theory enclosing Banach-spaces. The problems are presented in such a way that the theory can be generalized to apply to topological vector spaces.

Many different concrete programming problems possess one and the same theoretical basis, which we can formulate as a principle of duality. Many authors have generalized the well known relations of duality of linear programming for problems of non-linear programming. In this connection, the Fenchel theory of conjugate functions, and the relations of duality following on from this theory, plays an important role.

The main theme of this book is to represent such relations of duality and to consider a general theoretical basis for different special programming problems. In so doing, we do not claim to cover all programming problems but rather discuss some of the main ones by concentrating our enquiries on specially chosen Lagrangian forms. For linear programming we extend our investigations by constructing the theory in Banach spaces. In infinite dimensional Banach spaces we give examples to demonstrate the difference between finite dimensional and infinite dimensional theory and give a generalization of the transportation- and potential problem in finite directed graphs. We are not concerned here with constructions of algorithms for calculation of solutions.

In the first chapter the reader will find a collection of relations between convex sets, hyperplanes and extremal points which are required in the following chapters. Furthermore, we construct relations between convex functions and convex sets and describe the Fenchel theory of conjugate functions. In the second chapter, we formulate relations of duality and prove equivalence theorems with respect to saddle-point, duality and minimax theorems. Special structures of equivalence statements are very useful for well known linear programming problems to be embedded in. The third chapter deals with the embedding of different programming problems in connection with specially chosen Lagrangian forms and in addition to equivalence theorems, the chapter also contains existence theorems.

The basic conception of this book is an extended version of a course which the author gave at the Rangoon Arts and Science University during 1974/1975. The main sources were the University text the author prepared based on this course with the assistance of Professor Chit Swe.

M. Walk

Contents

1. Convex sets and convex functions

1.1. Convex sets

In the theory of optimization, the idea of convexity plays a central role. We shall begin our study of this concept with convex sets and their properties. In this monograph, we would restrict our discussions to real finite dimensional vector spaces, even though it is possible to extend most of the ideas and theorems to an abstract topological vector space. Inquiries of the following kind, one may find in all standard books on mathematical programming, especially works by VOGEL [32], KARLIN [20], BERGE and GHOUILA-HOURI [4], and KREKO [23].

Let $\mathbb{R}^n$ be a vector space of dimension n.

Definition 1.1.1. A set $C \subseteqq \mathbb{R}^n$ is *convex*, if for any two elements $u, v \in C$ and any real number λ, $0 \leqq \lambda \leqq 1$, we have

$$x = \lambda u + (1 - \lambda) v \in C. \tag{1.1.1}$$

An element $x = \sum_{i=1}^{r} \lambda_i x^i$ is a *convex combination* of vectors $x^1, x^2, \ldots, x^r \in \mathbb{R}^n$ if the real numbers $\lambda_1, \lambda_2, \ldots, \lambda_r$ are nonnegative and $\sum_{i=1}^{r} \lambda_i = 1$.

Theorem 1.1.1. *Let $C \subseteqq \mathbb{R}^n$ be a convex set. Every convex combination of vectors of C is an element of C.*

Proof: For $r = 1$, the assertion of the theorem is trivial.
Assume that the theorem is true for the value $r - 1$ with $r > 1$. Now consider vectors $x^1, x^2, \ldots, x^r \in C$ and nonnegative numbers $\lambda_1, \lambda_2, \ldots, \lambda_r$,where $\sum_{i=1}^{r} \lambda_i = 1$. If $\sum_{i=1}^{r-1} \lambda_i = 0$, then

$$x = \sum_{i=1}^{r} \lambda_i x^i = x^r$$

which is an element of C. If $\lambda = \sum_{i=1}^{r-1} \lambda_i \neq 0$ then we have

$$x = \sum_{i=1}^{r} \lambda_i x^i = \lambda \sum_{i=1}^{r-1} \tilde{\lambda}_i x^i + \lambda_r x^r$$

with $\tilde{\lambda}_i = \frac{\lambda_i}{\lambda}$, $i = 1, 2, \ldots, r - 1$. Obviously $\tilde{\lambda}_i \geqq 0$ and $\sum_{i=1}^{r-1} \tilde{\lambda}_i = 1$. By induction hypothesis $\tilde{x} = \sum_{i=1}^{r-1} \tilde{\lambda}_i x^i \in C$. Moreover, $\sum_{i=1}^{r} \lambda_i = 1$ implies $0 < \lambda \leqq 1$ and $\lambda_r = 1 - \lambda$. Thus, by Definition 1.1.1 $x = \lambda \tilde{x} + \lambda_r x^r \in C$. □

Theorem 1.1.2. *Let C_1 and C_2 be convex sets. If σ_1 and σ_2 are any two real numbers, then the set $C = \sigma_1 C_1 + \sigma_2 C_2 = \{x \in \mathbb{R}^n : x = \sigma_1 x^2 + \sigma_2 x^2, x^i \in C_i\}$ is also convex.*

Proof: Consider any two elements $u, v \in C$. Then there exist elements $u^1, v^1 \in C_1$ and corresponding elements $u^2, v^2 \in C_2$ such that

$$u = \sigma_1 u^1 + \sigma_2 u^2, \quad v = \sigma_1 v^1 + \sigma_2 v^2.$$

Now, for any real number λ, $0 \leqq \lambda \leqq 1$,

$$\begin{aligned} x = \lambda u + (1 - \lambda)\, v &= \lambda(\sigma_1 u^1 + \sigma_2 u^2) + (1 - \lambda)\,(\sigma_1 v^1 + \sigma_2 v^2) \\ &= \sigma_1\big(\lambda u^1 + (1 - \lambda)\, v_1\big) + \sigma_2\big(\lambda u^2 + (1 - \lambda)\, v^2\big) = \sigma_1 x^1 + \sigma_2 x^2 \end{aligned}$$

where x^1, x^2 are elements of convex sets C_1 and C_2. Consequently, $x \in C$. □

Regarding Theorem 1.1.2 we notice the following special cases:

(i) *If C_1 is a convex set and x^0 any element of $\mathbb{R}^n$, then the set*

$$C = C_1 + \{x^0\} = \{x \in \mathbb{R}^n : x = x^1 + x^0, x^1 \in C_1\}$$

is also convex.

(ii) *If C_1 is a convex set and σ any real number, then the set*

$$C = \sigma C_1 = \{x \in \mathbb{R}^n : x = \sigma x^1, x^1 \in C_1\}$$

is also convex.

(iii) *If C_1 is a convex set, then the set*

$$C = -C_1 = \{x \in \mathbb{R}^n : x = -x^1, x^1 \in C_1\}$$

is also convex.

Definition 1.1.2. Given $A \subseteqq \mathbb{R}^n$,

$$\operatorname{Conv}(A) = \left\{x \in \mathbb{R}^n : x = \sum_{i=1}^{r} \lambda_i x^i,\ x^i \in A,\ \lambda_i \geqq 0,\ \sum_{i=1}^{r} \lambda_i = 1,\ r = 1, 2, \ldots\right\}; \tag{1.1.2}$$

is called the *convex hull* of A.

Obviously Conv (A) is a closed set if the set A is finite. In case the set A is infinite, the convex hull Conv (A) is generally not closed. We may consider, for example,

$$A = \left\{\frac{1}{n} : n = 1, 2, \ldots\right\} \subseteqq \mathbb{R}.$$

One can easily find

$$\operatorname{Conv}(A) = \{x \in \mathbb{R} : 0 < x \leqq 1\}$$

which is not closed.

Theorem 1.1.3. *The convex hull* Conv (A) *of a set A is the smallest convex set, containing the set A.*

Proof: We shall first show that Conv (A) is convex. Consider $u, v \in$ Conv (A). Then u and v may be represented in the following forms:

$$u = \sum_{i=1}^{r} \varrho_i u^i, \quad u^i \in A, \quad \varrho_i \geqq 0, \quad \sum_{i=1}^{r} \varrho_i = 1,$$

$$v = \sum_{i=1}^{s} \sigma_i v^i, \quad v^i \in A, \sigma_i \geqq 0, \quad \sum_{i=1}^{s} \sigma_i = 1.$$

Now, for any real number λ, $0 \leq \lambda \leq 1$,

$$\begin{aligned} x &= \lambda u + (1 - \lambda)\, v \\ &= \sum_{i=1}^{r} \lambda \varrho_i u^i + \sum_{i=1}^{s} (1 - \lambda)\, \sigma_i v^i = \sum_{k=1}^{r+s} \tau_k z^k \end{aligned}$$

where

$$\tau_k = \begin{cases} \lambda \varrho_k, & k = 1, \ldots, r, \\ (1 - \lambda)\, \sigma_{k-r}, & k = r + 1, r + 2, \ldots, r + s, \end{cases}$$

$$z^k = \begin{cases} u^k \quad, & k = 1, \ldots, r, \\ v^{k-r}, & k = r + 1, \ldots, r + s. \end{cases}$$

We can observe that $z^k \in A$, $\tau_k \geqq 0$, $k = 1, \ldots, r + s$ and

$$\sum_{k=1}^{r+s} \tau_k = \lambda \sum_{k=1}^{r} \varrho_k + (1 - \lambda) \sum_{k=r+1}^{r+s} \sigma_{k-r} = \lambda + (1 - \lambda) = 1.$$

Thus $x \in$ Conv (A) implying that Conv (A) is convex. To complete the proof, we still have to show that for every convex set K containing A, Conv $(A) \subseteqq K$. This is quite clear because K is convex and from $A \subseteqq K$, it follows by Theorem 1.1.1 that every convex combination of elements of A belongs to K. □

The converse of Theorem 1.1.3 now follows:

Theorem 1.1.4. *If L is the smallest convex set containing A, then*

$$L = \text{Conv}\,(A).$$

Proof: L has the following properties:

(i) $A \subseteqq L$.

(ii) $L \subseteqq K$ for every convex set K containing A

Since Conv (A) is a convex set containing A, we have by (ii),

$$L \subseteqq \text{Conv}\,(A).$$

On the other hand, Theorem 1.1.3 together with (i) implies that

$$\text{Conv}\,(A) \subseteqq L.$$

Hence the proof of the theorem. □

Theorem 1.1.5. *The convex hull of a set A is the intersection of all convex sets which contain A:*

$$\operatorname{Conv}(A) = \bigcap_{A \subset K} K .$$

K being convex and containing A.

Proof. It is easily seen that the intersection of any collection of convex sets is convex. Thus, the set $\bigcap_{A \subset K} K$ is convex and it contains A. Furthermore, for any convex set K containing A, we have $\bigcap_{A \subset K} K$, i.e. $\bigcap_{A \subset K} K$ is the smallest convex set which contains A.

Now, by Theorem 1.1.4,

$$\bigcap_{A \subset K} K = \operatorname{Conv}(A). \square$$

The Theorems 1.1.1 to 1.1.5 are given only on the basis of the Definition 1.1.1 of convex sets. However, the proofs do not depend on the fact that we are restricting our considerations to finite dimensional Euclidean spaces. We should notice that if the discussions were in the more general framework of any linear space, the Theorems 1.1.1 to 1.1.5 are still true.

Besides convexity the following theorems are concerned with boundedness in finite dimensional spaces.

We will first deal with a theorem which says that in the case of an n-dimensional Euclidean space, the convex hull Conv (A) can be generated by all convex linear combinations of $n + 1$ or fewer elements of A.

Theorem 1.1.6. *If A is a subset of $\mathbb{R}^n$, then $x \in \operatorname{Conv}(A)$ can be represented by*

$$x = \sum_{i=1}^{r} \lambda_i x^i, \quad x^i \in A, \quad \lambda_i > 0, \quad \sum_{i=1}^{r} \lambda_i = 1, \quad r \leqq n + 1 .$$

Proof: For any $x \in \operatorname{Conv}(A)$,

$$x = \sum_{i=1}^{r} \lambda_i x^i, \quad x^i \in A, \quad \lambda_i \geqq 0, \quad \sum_{i=1}^{} \lambda_i = 1 .$$

We can assume $\lambda_i > 0$. If there is one $\lambda_i = 0$, we will drop this term off. If $r \leqq n + 1$, then there is nothing to prove. Now, consider the case $r > n + 1$. In this case, we can construct a new representation for x in which at most $r - 1$ elements of A appear. Such a construction is possible, because in an n-dimensional vector space, $n + 1$ vectors are always linearly dependent.

Let us consider vectors $z^i = x^i - x^r$, $i = 1, 2, \ldots, r - 1$. Under the assumption $r > n + 1$, the number of vectors z^i is more than n and in consequence the vectors z^i are linearly dependent. Thus there exist real numbers σ_i, $i = 1, \ldots, r - 1$, which are not all zero such that

$$\sum_{i=1}^{r-1} \sigma_i z^i = 0 .$$

Putting

$$\sigma_r = -\sum_{i=1}^{r-1} \sigma_i$$

we get

$$0 = \sum_{i=1}^{r-1} \sigma_i z^i = \sum_{i=1}^{r-1} \sigma_i x^i - x^r \sum_{i=1}^{r-1} \sigma_i = \sum_{i=1}^{r} \sigma_i x^i$$

and

$$\sum_{i=1}^{r} \sigma_i = 0 .$$

Now, for any $\varepsilon > 0$

$$x = \sum_{i=1}^{r} \lambda_i x^i = \sum_{i=1}^{r} (\lambda_i - \varepsilon\sigma_i)\, x^i . \tag{1.1.3}$$

Since at least one σ_i is positive, it is possible to calculate

$$\min \left\{ \frac{\lambda_i}{\sigma_i} : \sigma_i > 0 \right\} = \frac{\lambda_{i_0}}{\sigma_{i_0}} .$$

Here the index i_0 may not be uniquely defined. Now, if we put $\varepsilon = \dfrac{\lambda_{i_0}}{\sigma_{i_0}}$, then

$$\lambda_{i_0} - \varepsilon\sigma_{i_0} = 0 ,$$

$$\lambda_i - \varepsilon\sigma_i \neq 0 , \quad \text{for} \quad i \neq i_0$$

and

$$\sum_{r=1}^{r} (\lambda_\iota - \varepsilon\sigma_i) = 1 .$$

Now, from (1.1.3), for $\varepsilon = \dfrac{\lambda_{i_0}}{\sigma_{i_0}}$, we get the new representation

$$x = \sum_{\substack{i=1 \\ i \neq i_0}}^{r} (\lambda_i - \varepsilon\sigma_i)\, x^i .$$

In this form the element x will be represented by a convex linear combination of at most $r - 1$ elements of A. This construction, we can continue till we get the representation of x as a convex linear combination of at most $n + 1$ elements of A. □

Theorem 1.1.7. *Every set $\{x^k : k = 1, \ldots, r\}$ of r elements from $\mathbb{R}^n$ where $r \geqq n + 2$ can be divided into two parts such that the convex hulls of these parts have a nonempty intersection.*

Proof: The proof is based on the fact that there always exist a nontrivial solution for a homogeneous system of $n + 1$ linear algebraic equations in r unknowns, if $r \geqq n + 2$. Consider the set $\{x^k : k = 1, \ldots, r\}$, $r \geqq n + 2$. Then, there exists a nontrivial solution $\lambda_1, \lambda_2, \ldots, \lambda_r$ of the system.

$$\sum_{k=1}^{r} \lambda_k x^k = 0 , \quad \sum_{k=1}^{r} \lambda_k = 0 . \tag{1.1.4}$$

Define the two sets of integers

$$I = \{k : \lambda_k > 0\} , \quad J = \{1, 2, \ldots, r\} \setminus I .$$

One can easily see that I and J are nonempty and because of (1.1.4), we find

$$\sum_{k \in I} \lambda_k = - \sum_{k \in J} \lambda_k > 0 .$$

Furthermore, we get from (1.1.4) an element z in the form

$$z = \frac{\sum_{k\in I} \lambda_k x^k}{\sum_{k\in I} \lambda_k} = \frac{\sum_{k\in J} -\lambda_k x^k}{\sum_{k\in J} -\lambda_k}. \tag{1.1.5}$$

Clearly,

$$\sigma_k = \frac{\lambda_k}{\sum_{k\in I} \lambda_k} > 0 \quad \text{if} \quad k \in I; \quad \sum_{k\in I} \sigma_k = 1,$$

$$\sigma_k = \frac{-\lambda_k}{\sum_{k\in J} -\lambda_k} \geqq 0 \quad \text{if} \quad k \in J; \quad \sum_{k\in J} \sigma_k = 1.$$

The equation (1.1.5) means that the element z belongs to the convex hull of $\{x^k : k \in I\}$ as well as the convex hull of $\{x^k : k \in J\}$.

Theorem 1.1.8. (Theorem of Helly). *Let C_i, $i = 1, 2, \ldots, r$, $r \geqq n + 1$ be a finite class of convex sets in $\mathbb{R}^n$. If the intersection of any $n + 1$ of the sets $C_1, C_2, \ldots, C_r$ is nonempty, then the intersection of all r sets is nonempty.*

Proof: If $r = n + 1$, the assertion of the theorem is trivial.

Now assume that the theorem is true for $r - 1 \geqq n + 1$. This means if any $n + 1$ of the sets $C_1, \ldots, C_r$ have a nonempty intersection, the intersection of any $r - 1$ of them always has a nonempty intersection.

Now construct the set

$$\{x^k : k = 1, 2, \ldots, r\} \tag{1.1.6}$$

with elements $x^k \in \bigcap_{\substack{i=1\\ i\neq k}}^{r} C_i$.

Let $\{x^k : k \in I\}$ and $\{x^j : j \in J\}$ be a partition of the set (1.1.6) as given in Theorem 1.1.7. Every element x^k with $k \in I$ belongs to all sets C_j, $j \in J$, because for any $k \in I$ and $j \in J$, we have $k \neq j$. Consequently, for the convex hull of the set $\{x^k : k \in I\}$ we have

$$\operatorname{Conv}(\{x^k : k \in I\}) \subseteqq \bigcap_{j\in J} C_j. \tag{1.1.7}$$

Similarly we find that every element x^j with $j \in J$ belongs to all sets C_k, $k \in I$, and we have

$$\operatorname{Conv}(\{x^j : j \in J\}) \subseteqq \bigcap_{k\in I} C_k. \tag{1.1.8}$$

Then it follows from Theorem 1.1.7 that

$$\operatorname{Conv}(\{x^k : k \in I\}) \cap \operatorname{Conv}(\{x^j : j \in J\}) \neq \emptyset.$$

Let

$$z \in \operatorname{Conv}(\{x^k : k \in I\}) \cap \operatorname{Conv}(\{x^j : j \in J\}).$$

Because of (1.1.7) and (1.1.8)

$$z \in \left(\bigcap_{k\in I} C_k\right) \cap \left(\bigcap_{j\in J} C_j\right) = \bigcap_{i=1}^{r} C_i$$

which proves the theorem. □

At last we will show that the convex hull of a compact set is compact. For this reason we need the following lemma:

Lemma 1.1.1. *If the sets* $A \subseteqq \mathbb{R}^n$ *and* $B \subseteqq \mathbb{R}^m$ *are compact, then the product*

$$A \times B = \{(x, y)\} : x \in A, y \in B\}$$

is compact.

Proof: In a finite dimensional Euclidean space, a set is compact if and only if it is closed and bounded. For further discussions in the monograph we use the Euclidean norm. The following discussion shows that $A \times B$ is closed and bounded.

A and B being compact there exist numbers α and β such that

$$\sum_{i=1}^{n} \xi_i^2 = \|x\|^2 \leqq \alpha \quad \text{for any} \quad x \in A,$$

$$\sum_{i=1}^{m} \eta_i^2 = \|y\|^2 \leqq \beta \quad \text{for any} \quad y \in B.$$

If we recognize that a norm for elements $(x, y) \in A \times B$ is

$$\|(x, y)\| = (\|x\|^2 + \|y\|^2)^{\frac{1}{2}},$$

we get for any element $(x, y) \in A \times B$

$$\|(x, y)\|^2 = \|x\|^2 + \|y\|^2 \leqq \alpha + \beta .$$

Thus $A \times B$ is bounded.

The product $A \times B$ is also closed. Suppose $(x^k, y^k) \in A \times B$ is a converging sequence with a limit (x, y) i.e.

$$\lim_{k \to \infty} \|(x^k, y^k) - (x, y)\| = \lim_{k \to \infty} (\|x^k - x\|^2 + \|y^k - y\|^2)^{\frac{1}{2}} = 0.$$

This means the sequences $\{x^k\}$ and $\{y^k\}$ converge respectively to the limits x and y. Since A and B are closed $(x, y) \in A \times B$. □

The Lemma 1.1.1 and the fact that a continuous image of a compact set is compact lead us to the following theorem:

Theorem 1.1.9. *Let* $A \subseteqq \mathbb{R}^n$ *be compact. Then the convex hull* Conv (A) *is compact.*

Proof: We can easily observe that

$$S = \left\{x \in \mathbb{R}^{n+1} : \sum_{i=1}^{n+1} \xi_i = 1, \xi_i \geqq 0\right\}$$

is compact. By Lemma 1.1.1, the set

$$Y = S \times \underbrace{A \times \cdots \times A}_{n+1}$$

is compact.

We assign to each point $y \in Y$, by a continuous mapping ψ: $Y \to \mathbb{R}^n$, the value

$$\psi(y) = \psi(x, a^1, \ldots, a^{n+1}) = \sum_{i=1}^{n+1} \xi_i a^i \in \mathbb{R}^n .$$

Then the image of the continuous mapping,

$$\{\psi(Y)\} = \left\{z \in \mathbb{R}^n : z = \sum_{i=1}^{n+1} \xi_i a^i,\ a^i \in A,\ \xi_i \geqq 0,\ \sum_{i=1}^{n+1} \xi_i = 1\right\}$$

is compact. By Theorem 1.1.6, we have $\{\psi(Y)\} = \operatorname{Conv}(A)$. Consequently $\operatorname{Conv}(A)$ is compact. □

1.2. Hyperplanes and separation theorems

We will now present some principal relations between hyperplanes and convex sets. In the theory of duality of convex programming, such relations in the form of separation theorems play a central role especially in the proof of existence of optimal solutions. Of course, we apply the separation theorems also to an immediate characterization of convex sets. It is in this context, we will find them in the following discussions.

Definition 1.2.1. Let γ be a real number and $a \in \mathbb{R}^n$ any non zero fixed vector. The set $H \subseteqq \mathbb{R}^n$ given by

$$H = \{x \in \mathbb{R}^n : \langle a, x\rangle - \gamma = 0\}$$

is a *hyperplane* in $\mathbb{R}^n$.

Here $\langle a, x\rangle = \sum_{i=1}^{n} \alpha_i \xi_i$ denotes the scalar product of vectors $a = (\alpha_1, \alpha_2, \ldots, \alpha_n)$ and $x = (\xi_1, \xi_2, \ldots, \xi_n)$. Further we notice that the Euclidean norm of vector x may be expressed in the form:

$$\|x\| = \left(\sum_{i=1}^{n} \xi_i^2\right)^{\frac{1}{2}} = (\langle x, x\rangle)^{\frac{1}{2}}.$$

For $\gamma = 0$, the set H will form a $(n - 1)$-dimensional linear subspace of $\mathbb{R}^n$; otherwise, we get a linear manifold which arises by parallel translation of the subspace.

Geometrically in $\mathbb{R}^3$ a linear subspace $\langle a, x\rangle = 0$ represents a plane through the origin whereas the linear manifold $\langle a, x\rangle - \gamma = 0$ still represents a plane obtained by parallel displacement.

To prove the separation theorems, we need the following lemma:

Lemma 1.2.1. *If a closed convex set C is not empty and zero is not in C, then there exist exactly one point $x^0 \in C$ such that*

$$\|x^0\| = \min\{\|x\| : x \in C\}.$$

Proof: The proof of this lemma is based on the following well known facts:

(i) A continuous function on a compact set always attains its minimum.
(ii) Compactness in $\mathbb{R}^n$ means being closed and bounded.

Under the assumption that C is convex, the minimal element x^0 is uniquely defined.

Let $S = \{x \in \mathbb{R}^n : \|x\| \leqq \varrho\}$ be a closed sphere with centre at the origin, radius $\varrho > 0$ being chosen sufficiently great so that the set $A = C \cap S$ is not empty. Since A is bounded and closed, A is compact. Consequently, for the continuous function $\varphi(x) = \|x\|$, a vector $x^0 \in A$ exists such that

$$\varphi(x^0) = \|x^0\| = \min\{\varphi(x) : x \in A\}.$$

Because zero does not belong to C, it cannot be an element of A and thus $\|x^0\| > 0$. Furthermore, $\|x^0\| < \varrho$ and $\|x\| > \varrho$ for all $x \in C$, which are not in S.

Hence a fortiori

$$\|x^0\| \leqq \|x\|$$

for any $x \in C$. Consequently,

$$\varphi(x^0) = \|x^0\| = \min \{\varphi(x) : x \in C\}. \tag{1.2.1}$$

To complete the proof, we have to show that x^0 is uniquely defined. Let us assume that this is not so, and $x^1 \in C$ is another element such that

$$\|x^0\| = \|x^1\| = \min \{\varphi(x) : x \in C\}$$

and $x^0 \neq x^1$. Convexity of C implies that $x^2 = \frac{1}{2}(x^0 + x^1) \in C$. Finally we get

$$\|x^2\|^2 = \frac{1}{4}\|x^0 + x^1\|^2 < \frac{1}{4}(\|x^0 + x^1\|^2 + \|x^0 - x^1\|^2)$$

$$= \frac{1}{2}(\|x^0\|^2 + \|x^1\|^2) = \|x^0\|^2.$$

This means $\|x^2\| < \|x^0\|$ in contradiction to (1.2.1). □

From Lemma 1.2.1, we get a first separation theorem which says that a closed convex set not containing the origin can be strictly separated from the origin by a hyperplane.

Theorem 1.2.1. *If the convex set C is not empty and if the closure $\bar{C}$ of C does not contain the origin, then there exists a vector $a \neq 0$, such that*

$$\langle a, x\rangle \geqq \|a\|^2 > 0 \tag{1.2.2}$$

for any $x \in \bar{C}$.

Proof: We notice that the vector a which satisfies the assertion of the theorem can be chosen from $\bar{C}$. By Lemma 1.2.1, there exists a uniquely defined element $x^0 \in \bar{C}$ such that

$$\|x^0\| = \min \{\|x\| : x \in \bar{C}\}.$$

Choose $a = x^0$ and consider any $x \in \bar{C}$ and $0 < \lambda \leqq 1$. Because $\bar{C}$ is convex,

$$\lambda x + (1 - \lambda)\, x^0 = x^0 + \lambda(x - x^0) \in \bar{C}.$$

x^0 being a minimal element,

$$\|x^0\|^2 \leqq \|x^0 + \lambda(x - x_0)\|^2 = \|x^0\|^2 + 2\lambda\langle x^0, x - x^0\rangle + \lambda^2\,\|x - x^0\|^2.$$

Dividing by 2λ,

$$\langle x^0, x\rangle = \langle a, x\rangle \geqq \|a\|^2 - \frac{\lambda}{2}\|x - a\|^2.$$

Now, by letting λ converge to zero, we get the relation $\langle a, x\rangle \geqq \|a\|^2 > 0$ for any $x \in \bar{C}$. □

Theorem 1.2.1 can be generalized in a certain sense. Instead of a strict separation of a convex set from the origin, we consider two nonempty convex sets which are separated from each other by a positive distance and show that these two convex sets are separated strictly by a suitable hyperplane.

Definition 1.2.2. Let C_1 and C_2 be two nonempty convex sets. Then

$$\delta(C_1, C_2) = \inf \{\|x - y\|; x \in C_1, y \in C_2\}$$

denotes the *distance* between C_1 and C_2.

Theorem 1.2.2. *If C_1 and C_2 are two nonempty convex sets and the distance $\delta(C_1, C_2)$ is positive, then there exists a hyperplane*

$$H = \{x \in \mathbb{R}^n : \langle a, x\rangle - \gamma = 0\}$$

such that

$$\begin{aligned} &\langle a, x\rangle - \gamma > 0 \quad \textit{for any } x \in C_1, \\ &\langle a, x\rangle - \gamma < 0 \quad \textit{for any } x \in C_2. \end{aligned} \tag{1.2.3}$$

Proof: $\delta(C_1, C_2) > 0$ implies that the set $C = C_1 - C_2$ has a positive distance from zero i.e. the closure $\bar{C}$ does not contain zero. Since $C = C_1 - C_2$ is also convex, by Theorem 1.2.1, it follows that there exists a vector $a \in \mathbb{R}^n$, $a \neq 0$, such that for any $z \in C$

$$\langle a, z\rangle \geqq \|a\|^2 > 0.$$

Here $z = x - y$, $x \in C_1$, $y \in C_2$ and we get

$$\langle a, x\rangle \geqq \langle a, y\rangle + \|a\|^2 \tag{1.2.4}$$

for any $x \in C_1$ and $y \in C_2$. Thus we have

$$\inf \{\langle a, x\rangle : x \in C_1\} \geqq \sup \{\langle a, y\rangle : y \in C_2\} + \|a\|^2$$

i.e.

$$\inf \{\langle a, x\rangle : x \in C_1\} > \sup \{\langle a, y\rangle : y \in C_2\}.$$

Let γ be such that

$$\inf \{\langle a, x\rangle : x \in C_1\} > \gamma > \sup \{\langle a, y\rangle : y \in C_2\}.$$

Then, we can see that

$$H = \{x \in \mathbb{R}^n : \langle a, x\rangle - \gamma = 0\}$$

is the required hyperplane, satisfying the conditions

$$\begin{aligned} &\langle a, x\rangle - \gamma > 0 \quad \text{for any } x \in C_1, \\ &\langle a, y\rangle - \gamma < 0 \quad \text{for any } y \in C_2. \ \square \end{aligned}$$

In fact existence of separating hyperplanes is assured under more stringent conditions than those given in Theorems 1.2.1 and 1.2.2. A convex set which contains the origin on the boundary or two convex sets having a zero distance between them and having at the most boundary points in common can be separated by a suitable hyperplane. But in such cases, we cannot expect a strict separation but the separating hyperplane contains the origin in the former case and boundary points of the two convex sets in the latter.

Theorem 1.2.3. *Let C be a nonempty convex set with zero on the boundary of C. Then there exists a vector $a \in \mathbb{R}^n$, $a \neq 0$, such that for any $x \in \bar{C}$*

$$\langle a, x\rangle \geqq 0 \tag{1.2.5}$$

[Notice that zero may not be a point of C, if C is an open set.]

Proof: The origin being on the boundary of C, there exists a sequence $\{y^k\}$, $y^k \notin \bar{C}$ such that

$$\lim_{k\to\infty} y^k = 0.$$

Obviously, the sets

$$C_k = \bar{C} - \{y^k\} = \{z \in \mathbb{R}^n : z = x - y^k, x \in \bar{C}\}$$

are convex by Theorem 1.1.2, special case (i). Furthermore, the sets C_k are closed and zero does not belong to C_k for $y^k \notin \bar{C}$. By Theorem 1.2.1, there exists a sequence of vectors $a^k \in \mathbb{R}^n$ with $a^k \neq 0$ such that for any $z \in C_k$, we have

$$\langle a^k, z\rangle \geqq \|a^k\|^2 > 0.$$

From this inequality, we have for any $x \in \bar{C}$.

$$\langle a^k, x\rangle > \langle a^k, y^k\rangle = \gamma_k.$$

Without any loss of generality assume $\|a^k\| = 1$, otherwise we may divide by $\|a^k\|$. By Bolzano-Weierstrass theorem, there does exist a subsequence $\{a^{k_j}\}$ such that

$$\lim_{j\to\infty} a^{k_j} = a \in \mathbb{R}^n$$

where $\|a\| = 1$. Furthermore, we have

$$|\gamma_{k_j}| = |\langle a^{a_j}, y^{k_j}\rangle| \leqq \|a^{k_j}\| \, \|y^{k_j}\| = \|y^{k_j}\|.$$

Consequently, the sequence of real numbers γ_{k_j} converges to zero as $j \to \infty$, for $\lim_{j\to\infty} y^{k_j} = 0$ implies $\lim_{j\to\infty} \|y^{k_j}\| = 0$. Then from

$$\langle a^{k_j}, x\rangle > \gamma_{k_j},$$

for any $x \in \bar{C}$, by taking the limit as $j \to \infty$ of both sides, the assertion of the theorem follows. □

Theorem 1.2.4. *If C_1 and C_2 are nonempty convex sets and zero is a point on the boundary of $C_1 - C_2$, then there exists a hyperplane*

$$H = \{x \in \mathbb{R}^n : \langle a, x\rangle - \gamma = 0\}$$

such that

$$\begin{aligned} \langle a, x\rangle - \gamma &\geqq 0 \quad \textit{for any } x \in \bar{C}_1, \\ \langle a, x\rangle - \gamma &\leqq 0 \quad \textit{for any } x \in \bar{C}_2. \end{aligned} \tag{1.2.6}$$

Proof: By Theorem 1.2.3 there exists a vector a with $\|a\| = 1$ such that for any $z \in C_1 - C_2$,

$$\langle a, z\rangle \geqq 0.$$

With $z = x - y$, this inequality is equivalent to

$$\langle a, x\rangle \geqq \langle a, y\rangle \tag{1.2.7}$$

for any $x \in C_1$ and $y \in C_2$. Thus the set $\{\langle a, x\rangle : x \in C_1\}$ is bounded below i.e. there exists $\gamma = \inf \{\langle a, x\rangle : x \in C_1\}$. If we use this specially defined real number γ and the relation (1.2.7), we get

$$\begin{aligned} \langle a, x\rangle - \gamma &\geqq 0 \quad \text{for any } x \in C_1, \\ \langle a, y\rangle - \gamma &\leqq 0 \quad \text{for any } y \in C_2. \end{aligned}$$

These inequalities are still true if we add to the sets C_1 and C_2 all accumulation points. Thus we prove the assertion (1.2.6) of the theorem. □

If we consider the special case of Theorem 1.2.4 using a single element set $C_1 = \{b\}$ which is contained in the boundary of the set C_2, we get in a natural way a tangent hyperplane.

Definition 1.2.2. Let C be a convex set and b a point on the boundary of C. The hyperplane

$$H = \{x \in \mathbb{R}^n : \langle a, x\rangle - \gamma = 0\}$$

is called a *tangent hyperplane of C through the point b*, if we have

$$\langle a, x\rangle - \gamma \leqq 0 \quad \text{for any } x \in \overline{C},$$

$$\langle a, b\rangle - \gamma = 0.$$

If the set C_1 in Theorem 1.2.4 is a single element set $\{b\}$, then this theorem is reduced to the following theorem:

Theorem 1.2.5. *If C is a convex set and b a point on the boundary of C, then there exists a tangent hyperplane on C through b.*

It is possible to give another characterization of a tangent hyperplane. Every hyperplane H defines a closed half space H^+, is given by

$$H^+ = \{x \in \mathbb{R}^n : \langle a, x\rangle - \gamma \leqq 0\} \subseteqq \mathbb{R}^n .$$

Theorem 1.2.6. *A hyperplane H is a tangent hyperplane on C through the point b if and only if $\overline{C} \subseteqq H^+$ and $b \in H$.*

Proof: If H is a tangent hyperplane on C through b, then immediately by Definition 1.2.2,

$$\overline{C} \subseteqq H^+ \quad \text{and} \quad b \in H .$$

Now consider a hyperplane H and $b \in \overline{C}$. Suppose $\overline{C} \subseteqq H^+$ and $b \in H$. Then, we have

$$\langle a, x\rangle - \gamma \leqq 0 \quad \text{for any } x \in \overline{C}$$

and

$$\langle a, b\rangle - \gamma = 0 .$$

To complete the proof, we must show that b is a point on the boundary of C. Since $b \in H$, we can find a sequence $z^k \in H^- = \{x \in \mathbb{R}^n : \langle a, x\rangle - \gamma > 0\}$ such that $\lim_{k\to\infty} \|z^k - b\| = 0$. Because $b \in \overline{C}$ and $z^k \notin \overline{C}$, the point b must be necessarily on the boundary of C. □

Regarding Theorem 1.2.5, we can say that through any boundary point of a convex set there exists one tangent hyperplane. Thus the idea of characterizing a convex set C with the set of all tangent hyperplanes on C arises. This possible description is given by the following theorem:

Theorem 1.2.7. *Let C be a nonempty convex set. Then*

$$\overline{C} = \cap H^+$$

where the intersection is taken over all half spaces H^+, associated with the tangent hyperplanes of C.

Proof: If H is any tangent hyperplane on C, then Theorem 1.2.6 implies the relation $\overline{C} \subseteqq H^+$. Consequently $\overline{C} \subseteqq \cap H^+$.

To prove $\bar{C} = \bigcap H^+$ we assume $\bigcap H^+ \not\subseteq \bar{C}$. Then there exists an element $y^0 \in \bigcap H^+$ such that $y^0 \notin \bar{C}$. Thus, zero is not an element of the set $\{y^0\} - \bar{C}$ and consequently $\{y^0\} - \bar{C}$ has a positive distance from zero. By Theorem 1.2.1, there exists a vector $a \in \{y^0\} - \bar{C}$, $a \neq 0$ such that, for any $z \in \{y^0\} - \bar{C}$, we have

$$\langle a, z\rangle \geqq \|a\|^2 > 0 .$$

Equivalently, with $z = y^0 - x$,

$$\langle a, x\rangle \leqq \langle a, y^0\rangle - \|a\|^2 \tag{1.2.8}$$

for any $x \in \bar{C}$.

Now, we put $\gamma = \langle a, y^0\rangle - \|a\|^2$ and consider the hyperplane

$$H_0 = \{x \in \mathbb{R}^n : \langle a, x\rangle - \gamma = 0\} .$$

We now show that H_0 is a tangent hyperplane on C, which implies $H_0^+ \supseteqq \bigcap H^+$. It follows obviously from (1.2.8) that $\bar{C} \subseteqq H_0^+$. Furthermore, if $a \in \{y^0\} - \bar{C}$ then $x^0 = y^0 - a \in \bar{C}$, and

$$\langle a, x^0\rangle = \langle a, y^0\rangle - \|a\|^2 = \gamma .$$

Consequently, x^0 must be an element of H_0. By Theorem 1.2.6, H_0 is a tangent hyperplane on C through x^0. Furthermore from

$$\langle a, y^0\rangle > \langle a, y^0\rangle - \|a\|^2$$

follows $y^0 \notin H_0^+$, contrary to our assumption. Thus the element y^0 cannot be contained in $\bigcap H^+$ and this completes the proof. □

The set of boundary points of the convex set C is called the *profile* of C, which we denote by $\mathring{C}$. Now for any $b \in \mathring{C}$, in view of Theorem 1.2.5, there exists a tangent hyperplane through b on C. This hyperplane depends on b and we represent it in the form

$$H_b = \{x \in \mathbb{R}^n : \langle a^b, x\rangle - \gamma_b = 0\} .$$

If we choose any real number γ such that $\gamma \geqq \gamma_b$, we get obviously for any $x \in \bar{C}$

$$\langle a^b, x\rangle - \gamma \leqq 0 .$$

It means that the convex set $\bar{C}$ is completely on one side of the hyperplane

$$H = \{x \in \mathbb{R}^n : \langle a^b, x\rangle - \gamma = 0\} .$$

Now for H, there is another form of representation. If we have $a^b = (\alpha_1^b, \ldots, \alpha_n^b)$ and if we assume $\alpha_n^b \neq 0$, then the vector $(-1, c^b) \in \mathbb{R}^n$ with

$$c^b = \left(\frac{\alpha_1^b}{-\alpha_n^b}, \ldots, \frac{\alpha_{n-1}^b}{-\alpha_n^b}\right) \in \mathbb{R}^{n-1}$$

defines the direction of the hyperplane H.

Furthermore, the real number $-\dfrac{\gamma}{\alpha_n^b} = \delta$ defines the intercept made by H on ξ_n-coordinate axis. Using these facts, we can express H in the form

$$H = H_{(\delta, c^b)} = \{(\xi_n, \bar{x}) \in \mathbb{R}^n : \langle(-1, c^b), (\xi_n, \bar{x})\rangle - \delta = 0\} .$$

Here

$$\bar{x} = (\xi_1, \ldots, \xi_{n-1}) \in \mathbb{R}^{n-1}$$

and

$$\langle(-1, c^b), (\xi_n, \bar{x})\rangle = -\xi_n + \langle c^b, \bar{x}\rangle = -\xi_n + \sum_{i=1}^{n-1} \xi_i \frac{\alpha_i^b}{-\alpha_n^b}.$$

The hyperplane H is uniquely defined by the vector c^b and the intercept on the ξ_n-coordinate axis, $\delta = -\frac{\gamma}{\alpha_n^b}$. From this point of view, we can characterize the hyperplane H by an n-dimensional real vector $(\delta, c^b) \in \mathbb{R}^n$. Consequently, if $b \in \mathring{C}$, the vector

$$(\delta_b, c^b) = \left(\frac{\gamma_b}{-\alpha_n^b}, \frac{\alpha_1^b}{-\alpha_n^b}, \ldots, \frac{\alpha_{n-1}^b}{-\alpha_n^b}\right) \in \mathbb{R}^n$$

defines a tangent hyperplane through b on C in the form

$$H_b = H_{(\delta_b, c^b)} = \{(\xi_n, \bar{x}) \in \mathbb{R}^n : \langle(-1, c^b), (\xi_n, \bar{x})\rangle - \delta_b = 0\}.$$

Without any loss of generality we may assume $\alpha_n^b > 0$; it follows from $\gamma_b \leqq \gamma$, that for any δ with $\delta \leqq \delta_b$, a hyperplane defined by the vector (δ, c^b) has the convex set $\bar{C}$ completely on one side.

It means

$$\bar{C} \subseteqq H^+_{(\delta, c^b)}.$$

From now on, we will identify a hyperplane H with the vector $(\delta, c^b) \in \mathbb{R}^n$. Now we may assign to a convex set C a corresponding set $C^* \subseteqq \mathbb{R}^n$, defined by

$$C^* = \{(\delta, c^b) \in \mathbb{R}^n : b \in \mathring{C}; \delta \leqq \delta_b\}$$

and we call C^* the *conjugate* to C.

If we know the set C, the conjugate C^* will be uniquely defined and on the other hand, if we know the conjugate C^*, we can reproduce the convex set C. Obviously, regarding Theorem 1.2.7, we find the relation

$$\bar{C} = \bigcap_{b \in \mathring{C}} H^+_{(\delta_b, c^b)} = \bigcap_{(\delta, c^b) \in C^*} H^+_{(\delta, c^b)}.$$

For our later investigations, we need a theorem due to Berge [2], about intersections of convex sets, similar to the theorem of Helly.

Theorem 1.2.8 (Berge). *Let C_i, $i = 1, \ldots, r$ and $r \geqq 2$ be closed convex sets such that $\bigcup_{i=1}^{r} C_i$ is also convex. If all intersections of any $r - 1$ of these sets are not empty, it follows that the intersection of all r sets is not empty.*

Proof: We can confine our consideration to the case where the sets C_i are compact. If it is not so, we choose the points

$$a^i \in \bigcap_{j \neq i} C_j.$$

These points exist under the assumption of the theorem. The convex hull $\operatorname{Conv}(A) = \operatorname{Conv}(\{a^1, a^2, \ldots, a^r\})$ of a finite set A is a bounded closed polyhedron, which is compact. Because the sets C_i are closed, the convex sets $C_i' = \operatorname{Conv}(A) \cap C_i$ will be compact.

Furthermore, we have

$$C_i' \subseteqq C_i \quad \text{and} \quad a^i \in \bigcap_{j \neq i} C_j'.$$

If now the theorem holds for C_i', it also holds obviously with respect to the sets C_i. Hence the justification for the assumption that the sets C_i are compact.

The theorem may now be proved by induction. The assertion is true for the case $r = 2$, viz. if $C_1 \neq \emptyset$ and $C_2 \neq \emptyset$ and $C_1 \cup C_2$ is convex, then C_1 and C_2 cannot be disjoint. Otherwise by Theorem 1.2.2, we could have C_1 and C_2 strictly separated by a hyperplane H. Consequently there must be points of $C_1 \cup C_2$ on both sides of H and because $C_1 \cup C_2$ is convex, there even exist points in $C_1 \cup C_2$, which belong to H, strict separation of C_1 and C_2 implies that this is impossible.

Let the theorem be true for $r = m$ and $m \geqq 2$. Consider the sets $C_1, \ldots, C_{m+1}$ and put

$$C = \bigcap_{i=1}^{m} C_i .$$

Under the assumption of the theorem, C as well as C_{m+1} are not empty. If C and C_{m+1} are disjoint, we could separate these sets strictly by a hyperplane H.

Now, put

$$C_i' = H \cap C_i, \quad i = 1, \ldots, m .$$

Since $H \cap C_{m+1}$ is empty, we get

$$\bigcup_{i=1}^{m} C_i' = H \cap \left(\bigcup_{i=1}^{m} C_i \right) = H \cap \left(\bigcup_{i=1}^{m+1} C_i \right).$$

Consequently, $\bigcup_{i=1}^{m} C'$ is convex.

Now, we will show that any $(m - 1)$ sets from $C_1', \ldots, C_m'$ have a nonempty intersection. Putting

$$K_j = \bigcap_{\substack{i=1 \\ i \neq j}}^{m} C_i$$

we have $C \subseteqq K_j$ and $K_j \cap C_{m+1} \neq \emptyset$, because any m sets of $C_1, C_2, \ldots, C_{m+1}$ have a nonempty intersection.

This fact and the assumption that H separates C from C_{m+1} imply that the set $K_j \cap H$ is not empty.

It follows that

$$\bigcap_{\substack{i=1 \\ i \neq j}}^{m} C_i' = \bigcap_{\substack{i=1 \\ i \neq j}}^{m} (H \cap C_i) = H \cap \left(\bigcap_{\substack{i=1 \\ i \neq j}}^{m} C_i \right) = H \cap K_j \neq \emptyset .$$

Under the assumption of induction,

$$H \cap C = H \cap \left(\bigcap_{i=1}^{m} C_i \right) = \bigcap_{i=1}^{m} (H \cap C_i) = \bigcap_{i=1}^{m} C_i' \neq \emptyset$$

which is impossible, because the hyperplane H separates C from C_{m+1} strictly. Thus, C and C_{m+1} cannot be disjoint, i.e.

$$\bigcap_{i=1}^{m+1} C_i \neq \emptyset .$$

This proves the theorem. □

From Theorem 1.2.8, we get two corollaries:

Corollary 1.2.1. *Let* $C_1, \ldots, C_r$, $r \geqq 2$, *be closed convex sets. If intersections of* $r - 1$ *of these sets are not empty but the set* $\bigcap_{i=1}^{r} C_r$ *is empty, then* $\bigcap_{i=1}^{r} C_i$ *cannot be convex.*

This result follows immediately from the theorem.

Corollary 1.2.2. *Let* $C_1, \ldots, C_r$, $r \geqq 1$, *be* r *closed convex sets. If a convex set* C *is such that*

$$C \cap \left(\bigcap_{\substack{i=1 \\ i \neq j}}^{r} C_i \right) \neq \emptyset, \qquad j = 1, 2, \ldots, r,$$

but

$$C \cap \left(\bigcap_{i=1}^{r} C_i \right) = \emptyset,$$

then C *is not contained in* $\bigcup_{i=1}^{r} C_i$.

Proof: In the case where $r = 1$, the intersection of $(r - 1)$ sets being the whole space $\mathbb{R}^n$,

$$C \cap \mathbb{R}^n = C \neq \emptyset \quad \text{and} \quad C \cap C_1 = \emptyset$$

obviously implies that C is not contained in C_1. For the case where $r \geqq 2$, in view of the fact $C \cap \left(\bigcap_{\substack{i=1 \\ i \neq j}}^{r} C_i \right) \neq \emptyset$, we consider points $a^1, \ldots, a^r$ of C such that

$$a^j \in \bigcap_{\substack{i=1 \\ i \neq j}}^{r} C_i .$$

Put $A = \{a^1, \ldots, a^r\}$ and $C_i' = \operatorname{Conv}(A) \cap C_i$, $i = 1, 2, \ldots, r$. Now

$$a^j \in \operatorname{Conv}(A) \cap \left(\bigcap_{\substack{i=1 \\ i \neq j}}^{r} C_i \right) = \bigcap_{\substack{i=1 \\ i \neq j}}^{r} \left(\operatorname{Conv}(A) \cap C_i \right) = \bigcap_{\substack{i=1 \\ i \neq j}}^{r} C_i' \neq \emptyset .$$

Furthermore, regarding $C \cap \left(\bigcap_{i=1}^{r} C_i \right) = \emptyset$, we have,

$$\bigcap_{i=1}^{r} C_i' = \operatorname{Conv}(A) \cap \left(\bigcap_{i=1}^{r} C_i \right) = \emptyset .$$

Then, Corollary 1.2.1 implies that

$$\bigcup_{i=1}^{r} C_i' = \bigcup_{i=1}^{r} \operatorname{Conv}(A) \cap C_i = \operatorname{Conv}(A) \cap \left(\bigcup_{i=1}^{r} C_i \right)$$

is not convex. Consequently, there exist points $x^1, x^2 \in \bigcup_{i=1}^{r} C_i'$ such that for a real number λ, $0 \leqq \lambda \leqq 1$,

$$x = \lambda x^1 + (1 - \lambda)\, x^2 \notin \bigcup_{i=1}^{r} C_i' .$$

On the other hand, $x^1, x^2 \in \operatorname{Conv}(A)$ implies that $x \in \operatorname{Conv}(A)$. Obviously $x \notin \bigcup_{i=1}^{r} C_i$. But $x \in C$ for $\operatorname{Conv}(A) \subseteqq C$. Thus we have

$$C \not\subseteq \bigcup_{i=1}^{r} C_i . \square$$

1.3. Extremal points of convex sets

In problems of optimization where the set of points, in which we like to find an optimal solution, is a convex set, the extremal points of this set play a fundamental role. For example, there are well known results for certain problems of optimization which say that we can always realize a solution at an extremal point, if a solution exists. If the constraints form a polytope, with a finite number of extremal points, as we may find it in linear programming, we have to look for an optimal solution only among the finite set of extremal points and thus the problem becomes a discrete problem. In linear programming, this is the essential base which gives us the opportunity to construct the simplex algorithm. In this paragraph, we will make inquiries about relations between extremal points, convex sets and hyperplanes.

Let C be a convex set.

Definition 1.3.1. We call a point $x \in C$ an *extremal point* of C if from

$$x = \frac{1}{2}(y + z), \quad y \in C, \quad z \in C$$

follows

$$x = y = z.$$

If we consider for example the set $C = \{x \in \mathbb{R}^n : \|x\| \leqq 1\}$, then all the points $x \in C$ with $\|x\| = 1$ are extremal points. It means that the set C possesses a nondenumerably infinite number of extremal points. An equivalent characterization of extremal points is given by the following theorem.

Theorem 1.3.1. *A point $x \in C$ is an extremal point of C if and only if $x = \lambda y + (1 - \lambda) z$, with $0 < \lambda < 1$ and $y \in C$, $z \in C$ implies $x = y = z$.*

Proof: Let x be an extremal point and $x = \lambda y + (1 - \lambda) z$ with $0 < \lambda < 1$ and $y, z \in C$. Then, there exists an $\varepsilon > 0$ such that for $\lambda_1 = \lambda + \varepsilon$ and $\lambda_2 = \lambda - \varepsilon$, we have $0 \leqq \lambda_i \leqq 1$, $i = 1, 2$. Because C is convex $x^1 = \lambda_1 y + (1 - \lambda_1) z$ and $x^2 = \lambda_2 y + (1 - \lambda_2) z$ are elements of C if $y, z \in C$. We have

$$\frac{1}{2}(x^1 + x^2) = z + \frac{1}{2}(y - z)(\lambda_1 + \lambda_2) = z + \lambda(y - z) = x.$$

Since x is an extremal point, by Definition 1.3.1,

$$x = x^1 = x^2.$$

From this follows

$$(\lambda_1 - \lambda_2)(y - z) = 0.$$

Consequently $x = y = z$, because $\lambda_1 \neq \lambda_2$.

To prove sufficiently, we assume that from $x = \lambda y + (1 - \lambda) z$ with $0 < \lambda < 1$ and $y, z \in C$ follows $x = y = z$.

If we consider especially $\lambda = \frac{1}{2}$, x must be an extremal point by Definition 1.3.1. □

Remarks: Obviously it follows from Theorem 1.3.1. that $x \in C$ is an extremal point if and only if $x = \lambda y + (1 - \lambda) z$ with $0 \leqq \lambda \leqq 1$ and $y, z \in C$ with $y \neq z$ together implies that either $\lambda = 0$ or $\lambda = 1$.

If A is a finite subset of $\mathbb{R}^n$, the convex hull Conv (A) has at most a finite number of extremal points contained in A. This result follows from the following theorem:

Theorem 1.3.2. *Let E be the set of all extremal points of* Conv (A) *where A is a subset of $\mathbb{R}^n$. Then $E \subseteqq A$.*

Proof: Let x be an element of Conv (A) and $x \notin A$. Then for x, there exists a representation,

$$x = \sum_{i=1}^{k} \lambda_i x^i$$

with $x^i \in A$, $\lambda_i \geqq 0$ and $\sum_{i=1}^{k} \lambda_i = 1$. Because $x \notin A$, we must have $k > 1$. Without any loss of generality, we may assume

$$\lambda_k \neq 0 \quad \text{and} \quad \sum_{i=1}^{k-1} \lambda_i \neq 0.$$

Put

$$\lambda = \lambda_k, \quad 1 - \lambda = \sum_{i=1}^{k-1} \lambda_i$$

$$y = x^k, \quad z = \sum_{i=1}^{k-1} \frac{\lambda_i}{1 - \lambda} x^i.$$

Obviously $y, z \in$ Conv (A), $y \neq z$ and $0 < \lambda < 1$. Furthermore, we get

$$x = \lambda y + (1 - \lambda) z.$$

By Theorem 1.3.1, x cannot be an extremal point of Conv (A). Thus $E \subseteqq A$. □

Another characterization for an extremal point of a convex set now follows.

Theorem 1.3.3. *For a convex set C, a point $x \in C$ is an extremal point of C, if and only if $C \setminus \{x\}$ is convex.* Here $C \setminus \{x\}$ means all points from C except x.

Proof: Let x be an extremal point of C. Consider any two point $y, z \in C \setminus \{x\}$ and the point $u = \lambda y + (1 - \lambda) z$, with $0 \leqq \lambda \leqq 1$. If either $\lambda = 0$ or $\lambda = 1$, u ofcourse is an element of $C \setminus \{x\}$. For $0 < \lambda < 1$, assume that u is not an element of $C \setminus \{x\}$. However, because C is convex, u must at least be an element of C; thus $u = x = \lambda y + (1 - \lambda) z$.

Furthermore $x \neq y$ and $x \neq z$ implies that by Theorem 1.3.1 x cannot be an extremal point. We thus have a contradiction. Consequently $u \in C \setminus \{x\}$ so that $C \setminus \{x\}$ is convex.

Now, let $C \setminus \{x\}$ be a convex set and $x = \frac{1}{2} (y + z)$ with $y, z \in C$. If we have $x \neq y$, $x \neq z$, then in view of the convexity of $C \setminus \{x\}$, the point x must be an element of $C \setminus \{x\}$. But this is impossible. Thus $x = y = z$ whereby x is an extremal point of C. □

The answer to the question of existence of extremal points is provided by the following theorem:

Theorem 1.3.4. *A nonempty compact convex set C has always at least one extremal point.*

Proof: Obviously the function $\varphi(x) = \|x\|$ is a continuous function defined on $\mathbb{R}^n$. Because C is compact it is closed and bounded. Since a continuous function on a bounded closed set always attains its maximum or minimum, there exists an element $x^0 \in C$ such that

$$\|x^0\| = \max\{\|x\|: x \in C\}. \tag{1.3.1}$$

To show that x^0 is an extremal point, we use the convexity of C.

Assume

$$x^0 = \frac{1}{2}(y + z) = z + \frac{1}{2}(y - z) \quad \text{with} \quad y, z \in C.$$

By (1.3.1),

$$\|z\|^2 \leqq \|x^0\|^2 = \|z\|^2 + \langle z, y - z\rangle + \frac{1}{4}\|y - z\|^2.$$

From this follows

$$4\langle z, z - y\rangle \leqq \|y - z\|^2.$$

By symmetry, we can interchange y and z to get

$$4\langle y, y - z\rangle \leqq \|y - z\|^2.$$

Adding both inequalities, we have

$$4(\langle z, z - y\rangle + \langle y, y - z\rangle) = 4\|y - z\|^2 \leqq 2\|y - z\|^2.$$

Thus,

$$\|y - z\| = 0 \quad \text{or} \quad y = z.$$

Because C is a convex set, it follows from the Definition 1.3.1 that x^0 must be an extremal point. □

We shall next consider a relation between a convex set, a tangent hyperplane and an extremal point.

Theorem 1.3.5. *If C is a nonempty compact convex set, then any tangent hyperplane on C contains at least one extremal point.*

Proof: Let $H = \{x \in \mathbb{R}^n: \langle a, x\rangle - \gamma = 0\}$ be a tangent hyperplane on C. The set $L = H \cap C$ is a nonempty convex set and because C is compact, L must be compact. From Theorem 1.3.4, it follows that L has at least one extremal point.

Now, we will show that any extremal point of L must be an extremal point of C. Namely, if x^0 is an extremal point of L and

$$x^0 = \frac{1}{2}(y + z) \quad \text{with} \quad y, z \in C,$$

then using the property $C \subseteqq H^+$, we have

$$\langle a, y\rangle - \gamma \leqq 0,$$
$$\langle a, z\rangle - \gamma \leqq 0.$$

Furthermore,

$$0 = \langle a, x^0 \rangle - \gamma = \frac{1}{2} (\langle a, y \rangle - \gamma) + \frac{1}{2} (\langle a, z \rangle - \gamma).$$

This equation is only possible when

$$\langle a, y \rangle - \gamma = 0,$$
$$\langle a, z \rangle - \gamma = 0.$$

Consequently y, z are points of the hyperplane H and thus $y, z \in L$. x^0 being an extremal point of L, we have $x = y = z$. From this it follows that x^0 is an extremal point of C. □

At last, we will prove the theorem of KREIN-MILMAN [22] which gives us the possibility to represent a convex set, using the set of their extremal points.

Theorem 1.3.6. (KREIN-MILMAN). *Let C be a nonempty convex compact set and E the set of all extremal points of C. Then*

$$\overline{\mathrm{Conv}\,(E)} = C.$$

Proof: Because $E \subseteqq C$, we have $\mathrm{Conv}\,(E) \subseteqq C$ and because C is closed we have $\overline{\mathrm{Conv}\,(E)} \subseteqq C$.

Let us assume $C \not\subseteq \overline{\mathrm{Conv}\,(E)}$.

Then there exists an element $x^0 \in C$ such that $x^0 \notin \overline{\mathrm{Conv}\,(E)}$.

Now by Theorem 1.2.2, there exists a hyperplane

$$H = \{x \in \mathbb{R}^n : \langle a, x \rangle - \gamma = 0\}$$

such that H separated x^0 strictly from $\overline{\mathrm{Conv}\,(E)}$. From that follows, for any $x \in \overline{\mathrm{Conv}\,(E)}$,

$$\langle a, x \rangle - \gamma < 0 \tag{1.3.2}$$

and

$$\langle a, x^0 \rangle - \gamma > 0.$$

Because C is compact, there exists

$$\alpha = \max \{\langle a, x \rangle : x \in C\} = \langle a, z^0 \rangle, \qquad z^0 \in C.$$

Then

$$\alpha = \langle a, z^0 \rangle \geqq \langle a, x^0 \rangle > \gamma. \tag{1.3.3}$$

Next consider the hyperplane

$$H_\alpha = \{x \in \mathbb{R}^n : \langle a, x \rangle - \alpha = 0\}.$$

We will show that H_α is a tangent hyperplane on C through z^0. In fact from (1.3.3) follows $z^0 \in H_\alpha$ and furthermore, obviously for any $x \in C$, $\langle a, x \rangle - \alpha \leqq 0$. Then, by Theorem 1.2.6, H_α is a tangent hyperplane. By Theorem 1.3.5 H_α contains at least one extremal point y^0 of C. Thus $y^0 \in E$ and $y^0 \in H_\alpha$. But $y^0 \in E$ implies $y^0 \in \overline{\mathrm{Conv}\,(E)}$. Then by (1.3.2) and (1.3.3), we have

$$\langle a, y^0 \rangle - \alpha \leqq \langle a, y^0 \rangle - \gamma < 0.$$

This contradicts the fact $y^0 \in H_\alpha$. We arrive at this contradiction, by using a separation theorem under the assumption $C \not\subseteq \overline{\mathrm{Conv}\,(E)}$. Consequently, $C \subseteqq \overline{\mathrm{Conv}\,(E)}$ and this proves the theorem. □

1.4. Convex functions

In the theory of nonlinear programming, convex and concave functions are of special interest. This is because convex (concave) functions defined on closed convex sets have the property that a local maximum (minimum) is always a global maximum (minimum). Moreover, if the functions are strictly convex (concave), then the global maximum (minimum) is uniquely defined. To find examples which demonstrate the use of these properties, we may refer to HADLEY [17]. Here our investigations concern mainly with relations between convex and concave functions, convex sets and differentiability.

Definition 1.4.1. Let $C \subseteqq \mathbb{R}^n$ be a convex set. We call a function f defined on C a *convex function*, if for any elements $x, y \in C$ and $0 \leqq \lambda \leqq 1$, we have

$$f\big(\lambda x + (1-\lambda)\, y\big) \leqq \lambda f(x) + (1-\lambda)\, f(y).$$

If f is a convex function, we call $-f$ a *concave function*. Linear functions, for example, are convex as well as concave. Let f be a convex function defined on the convex set C. If $z = \sum_{i=1}^{k} \lambda_i x^i$ is a convex combination of vectors from C, we have obviously

$$f(z) = f\left(\sum_{i=1}^{k} \lambda_i x^i\right) \leqq \sum_{i=1}^{k} \lambda_i f(x^i).$$

This we can easily prove by induction. Other property of convex functions is given by the following theorem:

Theorem 1.4.1. *Let f be a convex function defined on a convex set C. If x^0 is an interior point of C, then f is continuous a x^0.*

Proof: Because x^0 is an interior point of C, there exists a number $\sigma > 0$ such that for any $y \in \mathbb{R}^n$ with $\|y\| < \sigma$ we have $x^0 + y \in C$. Now, we consider the function

$$g(y) = f(x^0 + y) - f(x^0)$$

defined on $S_\sigma = \{y \in \mathbb{R}^n : \|y\| < \sigma\}$. Obviously, the function g is continuous at zero if and only if f is continuous at x^0. Next consider a hypercube

$$Q = \{y = (\eta_1, \ldots, \eta_n) \in \mathbb{R}^n : |\eta_i| \leqq \varrho,\ i = 1, 2, \ldots, n\}$$

and choose the number ϱ such that Q is completely contained in S_σ. The hypercube Q is a convex compact set and has a finite number of extremal points. By Theorem 1.3.6 (KREIN-MILMAN), we can represent Q in the form of closure of convex hull of the extremal points $E = \{e^1, \ldots, e^k\}$ from Q. It then follows that for any element $y \in Q$, there exist real numbers $\lambda_1, \lambda_2, \ldots, \lambda_k$; $\lambda_i \geqq 0$ and $\sum_{i=1}^{k} \lambda_i = 1$ with $y = \sum_{i=1}^{k} \lambda_i e^i$. Now we get

$$\begin{aligned} g(y) &= f\left(x^0 + \sum_{i=1}^{k} \lambda_i e^i\right) - f(x^0) \\ &= f\left(\sum_{i=1}^{k} \lambda_i (x^0 + e^i)\right) - \sum_{i=1}^{k} \lambda_i f(x^0) \\ &\leqq \sum_{i=1}^{k} \lambda_i \big(f(x^0 + e^i) - f(x^0)\big) = \sum_{i=1}^{k} \lambda_i g(e^i). \end{aligned}$$

Thus we can estimate the function g from above, where the estimate is defined by the values of g at the extremal points of Q.

Now, choose

$$\alpha = \begin{cases} \max g(e^i),\ i = 1, \ldots, k \text{ in case there exist one } i \text{ with } g(e^i) > 0 \\ 1 \text{ in case } g(e^i) \leqq 0,\ i = 1, \ldots, k. \end{cases}$$

Obviously $g(y) \leqq \alpha$ for any $y \in Q$. Let $\varepsilon > 0$ be any real number such that $\frac{\varepsilon}{\alpha} \leqq 1$. Consider a $\delta(\varepsilon)$-neighborhood sphere of zero, $S_\delta = \{y \in \mathbb{R}^n : \|y\| \leqq \delta(\varepsilon)\}$ and put $\delta(\varepsilon) = \frac{\varepsilon\varrho}{\alpha}$. $\frac{\varepsilon}{\alpha} \leqq 1$ implies that S_δ is completely contained in Q. Moreover, we have for any $y \in S$, $\frac{\alpha}{\varepsilon} y \in Q$ and we get

$$g(y) = g\left(\frac{\varepsilon}{\alpha}\left[\frac{\alpha}{\varepsilon} y\right] + \left(1 - \frac{\varepsilon}{\alpha}\right) 0\right) \leqq \frac{\varepsilon}{\alpha} g\left(\frac{\alpha}{\varepsilon} y\right) + \left(1 - \frac{\varepsilon}{\alpha}\right) g(0)$$

$$= \frac{\varepsilon}{\alpha} g\left(\frac{\alpha}{\varepsilon} y\right) \leqq \varepsilon .$$

Furthermore, we get for any $y \in S_\delta$, from $0 = g(0) \leqq \frac{1}{2}\big(g(y) + g(-y)\big)$, the relation $g(y) \geqq -g(-y) \geqq -\varepsilon$. All together, it means $|g(y)| \leqq \varepsilon$ for any $y \in S_\delta$, i.e. g is continuous at zero. □

We notice that a convex function must not necessarily be continuous on the boundary. For example, the function

$$f(x) = \begin{cases} 2 & \text{for } x = 0, \\ 1 & \text{for } x \in (0, 1] \end{cases}$$

is convex on $[0, 1]$, but not continuous at zero.

Now we will study some relations between convex functions and convex sets as we need them in the theory of convex programming. Let g be a mapping from $\mathbb{R}^n$ into $\mathbb{R}^m$, usually denoted by $g: \mathbb{R}^n \to \mathbb{R}^m$. Then g has the form of a vector function, where the components are real valued functions $g_1, \ldots, g_m$, i.e.

$$g(x) = \big(g_1(x), \ldots, g_m(x)\big) \in \mathbb{R}^m .$$

If the functions g_i, $i = 1, 2, \ldots, m$, are concave and all functions are defined on one and the same convex set C, we shall call g a *concave mapping*.

The following theorem shows the relation between a concave mapping g and a $(n + m)$-dimensional set associated with g.

Theorem 1.4.2. *Let g be a concave mapping, defined on a convex set $C \subseteq \mathbb{R}^n$, into $\mathbb{R}^m$. Then the set*

$$[g, C] = \{(a, x) \in \mathbb{R}^m \times C : a \leqq g(x)\}$$

is convex.

On the other hand, if the sets C and $[g, C]$ are convex, then the mapping g is concave.

Here, we understand $a \leqq g(x)$ as the relation

$$\alpha_i \leqq g_i(x), \quad i = 1, \ldots, m, \text{ if } a = (\alpha_1, \ldots, \alpha_m).$$

Proof: Let g be concave and $(a^i, x^i) \in [g, C]$, $i = 1, 2$. Then we have $a^i \leqq g(x^i)$, $i = 1, 2$. Choose any real number λ, $0 \leqq \lambda \leqq 1$, and consider the point

$$(a, x) = \lambda(a^1, x^1) + (1 - \lambda)(a^2, x^2) = (\lambda a^1 + (1 - \lambda) a^2, \lambda x^1 + (1 - \lambda) x^2).$$

Obviously, $x = \lambda x^1 + (1 - \lambda) x^2 \in C$, and because the mapping g is concave, and using the definition of $[g, C]$, we get

$$g(x) \geqq \lambda g(x^1) + (1 - \lambda) g(x^2) \geqq \lambda a^1 + (1 - \lambda) a^2 = a.$$

It then follows that (a, x) is an element of $[g, C]$, i.e. $[g, C]$ is convex.

Now assume that the sets $[g, C]$ and C are convex. If x^1, x^2 are elements of C, then obviously the points $(g(x^1), x^1), (g(x^2), x^2) \in \mathbb{R}^m \times C$ belong to $[g, C]$. Under the assumption that $[g, C]$ is convex, the element

$$\left(\lambda g(x^1) + (1 - \lambda) g(x^2), \lambda x^1 + (1 - \lambda) x^2\right)$$

belongs to $[g, C]$ for any λ with $0 \leqq \lambda \leqq 1$. Then, the definition of $[g, C]$ implies

$$\lambda g(x^1) + (1 - \lambda) g(x^2) \leqq g\left(\lambda x^1 + (1 - \lambda) x^2\right),$$

i.e. g is concave. □

Later we will be interested in the special case $m = 1$. In this case, our concave mapping is a real valued concave function defined on a convex set C. If we replace this function by substituting $g = -f$, we have a convex function f. Applying Theorem 1.4.2, we find that the $(n + 1)$-dimensional set,

$$\begin{aligned}[g, C] &= \{(\alpha, x) \in \mathbb{R} \times C \colon -\alpha \leqq -f(x)\} \\ &= \{(\alpha, x) \in \mathbb{R} \times C \colon \alpha \geqq f(x)\} = [f, C]\end{aligned}$$

is convex if the function f is convex; on the other hand if the sets $[f, C]$ and C are convex, then the function f is convex. There is an interesting property of concave mapping as it is explained in the following theorem:

Theorem 1.4.3. *Let g be a concave mapping defined on a convex set $C \subseteqq \mathbb{R}^n$ into $\mathbb{R}^m$. Then the set*

$$B = \{y \in \mathbb{R}^m \colon y \leqq g(x) \text{ for at least one } x \in C\}$$

is convex.

Proof: Let y^1, y^2 be elements of B. Then there exist elements $x^1, x^2 \in C$ such that

$$y^1 \leqq g(x^1) \quad \text{and} \quad y^2 \leqq g(x^2).$$

Then, for a real number λ, $0 \leqq \lambda \leqq 1$, we have

$$y = \lambda y^1 + (1 - \lambda) y^2 \leqq \lambda g(x^1) + (1 - \lambda) g(x^2) \leqq g\left(\lambda x^1 + (1 - \lambda) x^2\right).$$

Because of the convexity of C, $x = \lambda x^1 + (1 - \lambda) x^2 \in C$. Thus, $y \leqq g(x)$ for at least one $x \in C$. Consequently $y \in B$ and the set B is convex. □

We notice that we cannot reverse the assertion of Theorem 1.4.3, though we can do so in Theorem 1.4.2. This means if the sets B and C are convex, g must not necessarily be concave. Let us consider, for example, a mapping from $C \subseteqq \mathbb{R}$ into $\mathbb{R}$ of the form:

$$g(\xi) = \xi^3 \quad \text{with} \quad C = [-1, 1].$$

The set

$$B = \{\eta \in \mathbb{R} : \eta \leqq \xi^3 \text{ for at least one } \xi \in [-1, 1]\}$$

is convex. Namely, if $\eta_1, \eta_2 \in B$, there exist $\xi_1, \xi_2 \in [-1, 1]$ such that $\eta_1 \leqq \xi_1{}^3$ and $\eta_2 \leqq \xi_2{}^3$. From this follows

$$\eta = \lambda\eta_1 + (1 - \lambda)\,\eta_2 \leqq \lambda\xi_1{}^3 + (1 - \lambda)\,\xi_2{}^3 \leqq 1 = g(1).$$

Because $1 \in [-1, 1]$, we can claim that there is at least one $\xi \in [-1, 1]$ such that $\eta \leqq g(\xi)$.

Consequently, $\eta \in B$ and B is convex but the function $g(\xi)$ is obviously not concave on $[-1, 1]$.

Theorem 1.4.4. *Let f be a convex function defined on a closed convex set C. Then a relative minimum of f over C is a global minimum. If f is strictly convex, then the global minimum is uniquely defined.* □

Here we call f *strictly convex*, if for any $x^0, x^1 \in C$ and for all real number λ, $0 < \lambda < 1$, we have

$$f\big(\lambda x^0 + (1 - \lambda)\,x^1\big) < \lambda f(x^0) + (1 - \lambda)\,f(x^1).$$

Proof: We assume f has at $x^1 \in C$ a relative minimum and at $x^0 \in C$, a global minimum implying $f(x^0) < f(x^1)$. Under this assumption, we can construct as shown below a contradiction to the fact that f attains a relative minimum at x^1. For a real number λ, $0 \leqq \lambda \leqq 1$, we have

$$f\big(\lambda x^0 + (1 - \lambda)\,x^1\big) \leqq \lambda f(x^0) + (1 - \lambda)\,f(x^1).$$

From $f(x^0) < f(x^1)$, it follows that for any λ, $0 < \lambda < 1$,

$$f\big(\lambda x^0 + (1 - \lambda)\,x^1\big) < \lambda f(x^1) + (1 - \lambda)\,f(x^1) = f(x^1). \tag{1.4.1}$$

Now we construct an ε-neighborhood sphere of x^1 for $\varepsilon > 0$,

$$S_\varepsilon(x^1) = \{x \in \mathbb{R}^n : \|x - x^1\| < \varepsilon\}.$$

Then, every element $x = \lambda x^0 + (1 - \lambda)\,x^1$ belongs to $S_\varepsilon(x^1)$ if λ satisfies

$$0 < \lambda < \frac{\varepsilon}{\|x^0 - x^1\|}.$$

In view of the inequality (1.4.1), we have for all such elements,

$$f(x) < f(x^1)$$

and this contradicts the fact that f attains a relative minimum at x^1. If f is strictly convex and if f attains its global minimum $f(x^0) = f(x^1)$ at $x^0, x^1 \in C$, where $x^0 \neq x^1$ we get for $x = \frac{1}{2}(x^0 + x^1) \in C$

$$f(x) < \frac{1}{2} f(x^0) + \frac{1}{2} f(x^1) = f(x^0) = f(x^1).$$

This contradicts the assumption that at x^0 and x^1 f attains the global minimum. Thus the global minimum points is uniquely defined in case f is strictly convex. □

In the following theorems we are concerned with differentiability of convex functions. In preparation for this study we make a few general remarks on differentiability of real valued functions. Let f be a real valued function defined on $C \subseteq \mathbb{R}^n$. If x^0 is an interior point of C, then for a suitable radius $\varrho > 0$, the sphere

$$S_\varrho(x^0) = \{x \in \mathbb{R}^n : \|x - x_0\| < \varrho\}$$

is completely contained in C. If $s \in \mathbb{R}^n$ is a vector such that $\|s\| < \varrho$, then obviously the point $x = x^0 + s$ belongs to $S_\varrho(x^0)$.

Definition 1.4.2. A linear mapping l from $\mathbb{R}^n$ into $\mathbb{R}$ is called a *Fréchet differential* of the function f at x^0 in the direction of s, if there exists a function $a(x^0, s)$ such that

$$f(x^0 + s) = f(x^0) + l(s) + a(x^0, s)$$

with

$$\lim_{\|s\| \to 0} \frac{|a(x^0, s)|}{\|s\|} = 0.$$

The Fréchet differential depends on the point x^0.

By a *linear mapping* l we understand a continuous mapping with the property $l(\lambda_1 x^1 + \lambda_2 x^2) = \lambda_1 l(x^1) + \lambda_2 l(x^2)$. We notice that the Fréchet differential is a linear functional and it is a concept quite different from that of the derivatives of a function.

We sometimes use the denotation $(\mathrm{d}f)_{x^0} = l$ for the Fréchet differential.

If f possesses at x^0 a Fréchet differential, we say that f is *Fréchet differentiable* or F-*differentiable at* x^0. We observe that the Fréchet differential is uniquely defined.

Let l_1, l_2 be two Fréchet differentials of f at x^0. Then we have

$$f(x^0 + s) - f(x^0) = l_1(s) + a_1(x^0, s) = l_2(s) + a_2(x^0, s)$$

with

$$\lim_{\|s\| \to 0} \frac{|a_1(x^0, s)|}{\|s\|} = \lim_{\|s\| \to 0} \frac{|a_2(x^0, s)|}{\|s\|} = 0.$$

From this follows

$$\lim_{\|s\| \to 0} \frac{|(l_1 - l_2)(s)|}{\|s\|} = \lim_{\|s\| \to 0} \frac{|a_2(x^0, s) - a_1(x^0, s)|}{\|s\|} = 0.$$

If s^0 is any nonzero vector in $\mathbb{R}^n$ and if we put $s = \mu s^0$, where μ is a positive real number, then obviously $\lim\limits_{\mu \to 0} \|s\| = 0$. Thus we get

$$\lim_{\|s\| \to 0} \frac{|(l_1 - l_2)(s)|}{\|s\|} = \lim_{\mu \to 0} \left| (l_1 - l_2) \left(\frac{\mu s^0}{|\mu| \, \|s^0\|} \right) \right|$$
$$= \lim_{\mu \to 0} \left| (l_1 - l_2) \left(\operatorname{sgn} \mu \, \frac{s_0}{\|s^0\|} \right) \right| = 0.$$

Because $\mu > 0$ we have $\operatorname{sgn} \mu = 1$ and the limiting value does not depend on μ.

Consequently, we have, for any vector $s^0 \in \mathbb{R}^n$

$$\|s^0\|^{-1} (l_1 - l_2)(s^0) = 0.$$

This is possible only for $l_1 = l_2$.

As in the case of usual differentiability of functions, we find that Fréchet differentiability implies continuity.

Theorem 1.4.5. *If the function f is differentiable at x^0 in the sense of Fréchet, then f is continuous at x^0.*

Proof: Because the linear functional l is continuous, there exists for any $\varepsilon > 0$ a positive real number $\delta_1(\varepsilon)$ such that for all vectors s with $\|s\| < \delta_1(\varepsilon)$, we have

$$|l(s)| \leqq \frac{\varepsilon}{2}.$$

Furthermore, since $\dfrac{|a(x^0, s)|}{\|s\|}$ together with $\|s\|$ tends to zero, there exists a positive real number $\delta_2(\varepsilon)$ such that for all s with $\|s\| < \delta_2(\varepsilon)$, we have

$$|a(x^0, s)| \leqq \frac{\varepsilon}{2}.$$

Now we put $\delta = \min\{\delta_1, \delta_2\}$. Then for all s with $\|s\| < \delta$,

$$|f(x^0 + s) - f(x^0)| \leqq |l(s)| + |a(x^0, s)| \leqq \varepsilon,$$

and this proves the theorem. □

If $x = (\xi_1, \ldots, \xi_n) \in \mathbb{R}^n$, we denote by $D_i f(x) = \dfrac{\partial f}{\partial \xi_i}$, the partial derivative of f with respect to ξ_i. The relation between Fréchet differential and partial derivatives is now given.

Theorem 1.4.6. *If f is Fréchet differentiable at x^0 and $e^i = (0, \ldots, 1, \ldots, 0) \in \mathbb{R}^n$ is the ith basic vector, then there exist the partial derivatives $D_i f(x_0)$ of f at x^0 and*

$$l(e^i) = (\mathrm{d}f)_{x^0}(e^i) = D_i f(x^0).$$

Proof: Let $\alpha > 0$ be a real number. Then we have

$$\lim_{\alpha \to 0} \|\alpha e^i\| = \lim_{\alpha \to 0} |\alpha| = 0.$$

Because f is Fréchet differentiable, we get for $s = \alpha e^i$,

$$f(x^0 + \alpha e^i) - f(x^0) = l(\alpha e^i) + a(x^0, \alpha e^i)$$

with

$$\lim_{\alpha \to 0} \frac{|a(x^0, \alpha e^i)|}{|\alpha|} = 0.$$

From this follows

$$\lim_{\alpha \to 0} \left| \frac{f(x^0 + \alpha e^i) - f(x^0)}{\alpha} - l(e^i) \right| = \lim_{\alpha \to 0} \frac{|a(x^0, \alpha e^i)|}{|\alpha|} = 0,$$

which is equivalent to $l(e^i) = D_i f(x^0)$.

Let $s = (\sigma_1, \ldots, \sigma_n)$ be any vector in $\mathbb{R}^n$ so that $s = \sum_{i=1}^{n} \sigma_i e^i$. If f is Fréchet differentiable at x^0, we get

$$(\mathrm{d}f)_{x^0}(s) = (\mathrm{d}f)_{x^0}\left(\sum_{i=1}^{n} \sigma_i e^i\right) = \sum_{i=1}^{n} \sigma_i (\mathrm{d}f)_{x^0}(e^i).$$

From this follows, by Theorem 1.4.6, a representation for the Fréchet differential in the form,

$$(\mathrm{d}f)_{x^0}(s) = \sum_{i=1}^{n} \sigma_i \mathrm{D}_i f(x^0) = \langle \operatorname{grad} f(x^0), s \rangle.$$

Here we use the denotion

$$\operatorname{grad} f(x^0) = \big(\mathrm{D}_1 f(x^0), \ldots, \mathrm{D}_n f(x^0)\big) \in \mathbb{R}^n.$$

Thus $(\mathrm{d}f)_{x^0}(s) = \langle \operatorname{grad} f(x^0), s \rangle$ is the total differential of the function f.

Theorem 1.4.6 is not reversible in the sense that from the existence of first partial derivatives, we cannot deduce the existence of a Fréchet differential. We will now demonstrate this fact with an example.

Let us consider the case $n = 2$ and a function

$$f(x) = f(\xi_1, \xi_2) = \sqrt{|\xi_1 \xi_2|}$$

defined on $C = \mathbb{R}^2$. If $\xi_i \neq 0$, there exist the first partial derivatives

$$\mathrm{D}_1 f(x) = \frac{(\operatorname{sgn} \xi_1)(\operatorname{sgn} \xi_2)\, \xi_2}{2\sqrt{|\xi_1 \xi_2|}}, \qquad \mathrm{D}_2 f(x) = \frac{(\operatorname{sgn} \xi_1)(\operatorname{sgn} \xi_2)\, \xi_1}{2\sqrt{|\xi_1 \xi_2|}}.$$

At the origin, $x = (\xi_1, \xi_2) = (0, 0)$ the first partial derivatives have the values

$$\mathrm{D}_1 f(0) = \lim_{\alpha \to 0} \frac{f(\alpha, 0) - f(0, 0)}{\alpha} = \lim_{\alpha \to 0} \frac{0}{\alpha} = 0,$$

$$\mathrm{D}_2 f(0) = \lim_{\alpha \to 0} \frac{f(0, \alpha) - f(0, 0)}{\alpha} = \lim_{\alpha \to 0} \frac{0}{\alpha} = 0.$$

From this follows f has first partial derivatives at all points of $\mathbb{R}^2$. If f were Fréchet differentiable at $x = 0$, then according to the special representation of the differential in partial derivatives, we have $(\mathrm{d}f)_{x=0}(s) = \sigma_1 \mathrm{D}_1 f(0) + \sigma_2 \mathrm{D}_2 f(0) = 0$. Then, in view of Definition 1.4.2, we find

$$f(0 + s) - f(0) - (\mathrm{d}f)_0(s) = \sqrt{|\sigma_1 \sigma_2|} = a(0, s).$$

One can easily see that $\frac{|a(0, s)|}{\|s\|}$ does not converge to the value zero as $\|s\| \to 0$.

Along $\sigma_1 = \sigma_2$, we get $\lim\limits_{\sigma_1 = \sigma_2 \to 0} \frac{|a(0, s)|}{\|s\|} = \lim\limits_{\sigma_1 = \sigma_2 \to 0} \frac{\sqrt{\sigma_1^2}}{\sqrt{2}\sqrt{\sigma_1^2}} = \frac{1}{\sqrt{2}}$. On the other hand, $\lim\limits_{\sigma_2 \to 0} \lim\limits_{\sigma_1 \to 0} \frac{\sqrt{|\sigma_1 \sigma_2|}}{\sqrt{\sigma_1^2 + \sigma_2^2}} = \lim\limits_{\sigma_1 \to 0} \lim\limits_{\sigma_2 \to 0} \frac{\sqrt{|\sigma_1 \sigma_2|}}{\sqrt{\sigma_1^2 + \sigma_2^2}} = 0$. Consequently the Fréchet differential for f does not exist at $x = 0$. If we assume that the first partial derivatives of f are continuous then we can reverse the statement of Theorem 1.4.6.

Theorem 1.4.7. *If f possesses, at an interior point $x^0 \in C$, first continuous partial derivatives, then the Fréchet differential of f at x^0 exists and we have*

$$(\mathrm{d}f)_{x^0} = \operatorname{grad} f(x^0).$$

Proof: First we put $(\mathrm{d}f)_{x^0} = \operatorname{grad} f(x^0)$. Then $(\mathrm{d}f)_{x^0}$ is a continuous linear functional. Now we have

$$\lim_{\|s\|\to 0} \frac{|a(x^0, s)|}{\|s\|} = \lim_{\|s\|\to 0} \frac{f(x^0 + s) - f(x^0) - \langle \operatorname{grad} f(x^0), s\rangle}{\|s\|}.$$

We will show that this limit is zero with a view to proving the theorem. We have

$$\frac{1}{\|s\|}\left(f(x^0 + s) - f(x^0) - \sum_{i=1}^{n} \sigma_i \mathrm{D}_i f(x^0)\right)$$
$$= \frac{1}{\|s\|}\left(\sum_{i=1}^{n-1} [f(\xi_1^0 + \sigma_1, \ldots, \xi_{i+1}^0 + \sigma_{i+1}, \xi_{i+2}^0, \ldots, \xi_n^0)\right.$$
$$- f(\xi_1^0 + \sigma_1, \ldots, \xi_i^0 + \sigma_i, \xi_{i+1}^0, \ldots, \xi_n^0) - \sigma_{i+1}\mathrm{D}_{i+1} f(x^0)]$$
$$\left. + f(\xi_1^0 + \sigma_1, \xi_2^0, \ldots, \xi_n^0) - f(\xi_1^0, \ldots, \xi_n^0) - \sigma_1 \mathrm{D}_1 f(x^0)\right).$$

Furthermore we conclude

$$\frac{1}{\|s\|} = \frac{1}{\sqrt{\sigma_1^2 + \cdots + \sigma_n^2}} = \frac{1}{|\sigma_{i+1}|} \frac{1}{\sqrt{\frac{\sigma_1^2}{\sigma_{i+1}^2} + \cdots + 1 + \cdots + \frac{\sigma_n^2}{\sigma_{i+1}^2}}} = \frac{1}{|\sigma_{i+1}|} K_{i+1}$$

where $0 \leqq K_{i+1} \leqq 1$ and σ_{i+1} tends to zero as $\|s\| \to 0$. Then we get

$$\lim_{\|s\|\to 0} \frac{1}{\|s\|} \left(f(\xi_1^0 + \sigma_1, \xi_2^0, \ldots, \xi_n^0) - f(\xi_1^0, \xi_2^0, \ldots, \xi_n^0) - \sigma_1 \mathrm{D}_1 f(x^0)\right)$$
$$= \lim_{\|s\|\to 0} K_1 \operatorname{sgn} \sigma_1 \left(\frac{f(\xi_1^0 + \sigma_1, \xi_2^0, \ldots, \xi_n^0) - f(\xi_1^0, \ldots, \xi_n^0)}{\sigma_1} - \mathrm{D}_1 f(x^0)\right) = 0.$$

Furthermore,

$$\lim_{\|s\|\to 0} \frac{1}{\|s\|} \left(f(\xi_1^0 + \sigma_1, \ldots, \xi_{i+1}^0 + \sigma_{i+1}, \xi_{i+2}^0, \ldots, \xi_n^0)\right.$$
$$\left. - f(\xi_1^0 + \sigma_1, \ldots, \xi_i^0 + \sigma_i, \xi_{i+1}^0, \ldots, \xi_n^0) - \sigma_{i+1} \mathrm{D}_{i+1} f(x^0)\right)$$
$$= \lim_{\|s\|\to 0} K_{i+1} \operatorname{sgn} \sigma_{i+1} \left(\frac{f(\xi_1^0 + \sigma_1, \ldots, \xi_{i+1}^0 + \sigma_{i+1}, \xi_{i+2}^0, \ldots, \xi_n^0)}{\sigma_{i+1}}\right.$$
$$\left. - \frac{f(\xi_1^0 + \sigma_1, \ldots, \xi_i^0 + \sigma_i, \xi_{i+1}^0, \ldots, \xi_n^0)}{\sigma_{i+1}} - \mathrm{D}_{i+1} f(x^0)\right).$$

If we use mean value theorem, then

$$\frac{f(\xi_1^0 + \sigma_1, \ldots, \xi_{i+1}^0 + \sigma_{i+1}, \xi_{i+2}^0, \ldots, \xi_n^0) - f(\xi_1^0 + \sigma_1, \ldots, \xi_i^0 + \sigma_i, \xi_{i+1}^0, \ldots, \xi_n^0)}{\sigma_{i+1}}$$
$$= \mathrm{D}_{i+1} f(\xi_1^0 + \sigma_1, \ldots, \xi_{i+1}^0 + \theta\sigma_{i+1}, \xi_{i+2}^0, \ldots, \xi_n^0)$$

where θ is a suitable real number with $0 < \theta < 1$. Because the first derivatives of f are continuous, we get

$$\lim_{\|s\|\to 0} K_{i+1} \operatorname{sgn} \sigma_{i+1}\left(\mathrm{D}_{i+1}f(\xi_1^0 + \sigma_1, \ldots, \xi_{i+1}^0 + \theta\sigma_{i+1}, \xi_{i+2}^0, \ldots, \xi_n^0) - \mathrm{D}_{i+1}f(x^0)\right) = 0.$$

Altogether we have

$$\lim_{\|s\|\to 0} \frac{f(x^0 + s) - f(x^0) - \sum_{i=1}^{n} \sigma_i \mathrm{D}_i f(x^0)}{\|s\|} = 0$$

and this proves the theorem. □

The following theorem expresses a relation between Fréchet differential of a function at a point and a local maximum of this function.

Theorem 1.4.8. *If x^0 is an interior point of the set*

$$\mathbb{R}^{n+} = \{x \in \mathbb{R}^n : x \geqq 0\}$$

and if the Fréchet differentiable function f possesses a local maximum at x^0 then we have

$$(\mathrm{d}f)_{x^0} = \operatorname{grad} f(x^0) \leqq 0,$$

$$\langle (\mathrm{d}f)_{x^0}, x^0 \rangle = \langle \operatorname{grad} f(x^0), x^0 \rangle = 0.$$

Proof: Because f has at $x^0 \in \mathbb{R}^{n+}$ a local maximum, there exist an $\varepsilon > 0$ such that for any $x \in \mathbb{R}^{n+} \cap S_\varepsilon(x^0)$, we have

$$0 \geqq f(x) - f(x^0) = (\mathrm{d}f)_{x^0}\,(x - x^0) + a(x^0, x - x^0).$$

If we choose a real number η with $0 < \eta < \varepsilon$, then all points $x = x^0 + \eta e^i$ belong to $\mathbb{R}^{n+} \cap S_\varepsilon(x^0)$ and we get by substitution

$$0 \geqq \frac{f(x^0 + \eta e^i) - f(x^0)}{\eta} = (\mathrm{d}f)_{x^0}\,(e^i) + \frac{a(x^0, \eta e^i)}{\eta}.$$

On taking the limit as $\eta \to 0$

$$0 \geqq \mathrm{D}_i f(x^0) = (\mathrm{d}f)_{x^0}\,(e^i).$$

This inequality for $i = 1, 2, \ldots, n$ leads to

$$\operatorname{grad} f(x_0) \leqq 0$$

and this is the first part of the theorem. Next consider the scalar product

$$\langle \operatorname{grad} f(x^0), x^0 \rangle.$$

For $x^0 \in \mathbb{R}^{n+}$, we have $\xi_i^0 \geqq 0$, $i = 1, \ldots, n$, and only those strictly positive components of x^0 are of interest here.

If ξ_k^0 were strictly positive, then there exist a number $\eta > 0$ such that $x = x^0 - \eta e^k \geqq 0$. Furthermore, if $\eta \leqq \varepsilon$, we have

$$0 \geqq f(x^0 - \eta e^k) - f(x^0) = (\mathrm{d}f)_{x^0}\,(-\eta e^k) + a(x^0, -\eta e^k).$$

Dividing by η leads us to

$$0 \geqq \frac{f(x^0 - \eta e^k) - f(x^0)}{\eta} = -(\mathrm{d}f)_{x^0}(e^k) + \frac{a(x^0, -\eta e^k)}{\eta}.$$

From this inequality, it follows that, as $\eta \to 0$,

$$(\mathrm{d}f)_{x^0}(e^k) = \mathrm{D}_k f(x^0) \geqq 0.$$

Because, in general we have grad $f(x^0) \leqq 0$ implying $\mathrm{D}_k f(x^0) \leqq 0$ for all k, it follows that $\mathrm{D}_k f(x^0) = 0$ in case $\xi_0{}^k > 0$.

Altogether we have $\langle \operatorname{grad} f(x^0), x^0 \rangle = 0$ and this proves the theorem. □

We could see that from the existence of a Fréchet differential follows the existence of the first partial derivatives, but this is not reversible. Besides the existence of first partial derivatives, if we know that those derivatives are continuous, then Theorem 1.4.7 shows that the Fréchet differential is equal to the gradient of the function. If we replace the assumption that the first partial derivatives are continuous by an assumption that the function f is convex, then the existence of a Fréchet differential follows from the existence of first partial derivatives, as it is explained in the following theorem.

Theorem 1.4.9. *Let f be a convex function defined on a convex set C, of which x^0 is an interior point. If f possesses first partial derivatives at x^0 then the Fréchet differential of f exist and*

$$(\mathrm{d}f)_{x^0} = \operatorname{grad} f(x^0).$$

Proof: We put $f(x^0 + s) - f(x^0) - \langle \operatorname{grad} f(x^0), s \rangle = a(x^0, s)$ where $s \in \mathbb{R}^n$ belongs to a suitable ε-neighborhood sphere $S_\varepsilon(0)$ of zero such that $x^0 + s \in C$. Because f is a convex function, so also is the function $a(x^0, s)$ convex with respect to s in $S_\varepsilon(0)$.

From this we get for $s = \sum_{i=1}^{n} \sigma_i e^i$

$$a(x^0, s) = a\left(x^0, \sum_{i=1}^{n} \frac{1}{n}(n\sigma_i e^i)\right) \leqq \frac{1}{n} \sum_{i=1}^{n} a(x^0, n\sigma_i e^i) = \sum_{i=1}^{n} \sigma_i \chi_i$$

with

$$\chi_i = \frac{f(x^0 + n\sigma_i e^i) - f(x^0)}{n\sigma_i} - \mathrm{D}_i f(x^0).$$

Obviously χ_i converges to zero if $\|s\| \to 0$ because of the existence of the first partial derivatives of f. Furthermore, it follows from Schwarz inequality that

$$|a(x^0, s)| \leqq \left|\sum_{i=1}^{n} \sigma_i \chi_i\right| \leqq \|s\| \left(\sum_{i=1}^{n} |\chi_i|^2\right)^{\frac{1}{2}}.$$

From this follows

$$\lim_{\|s\| \to 0} \frac{|a(x^0, s)|}{\|s\|} = \lim_{\|s\| \to 0} \left(\sum_{i=1}^{n} |\chi_i|^2\right)^{\frac{1}{2}} = 0.$$

This proves the theorem. □

In our future inquiries, we find it useful to have the estimate of the change of value of a convex function with points of a neighborhood using the Fréchet differential. This relation is now given.

Theorem 1.4.10. *Let f be a convex function defined on a convex set C of which x^0 is an interior point. If f is Fréchet differentiable at x^0, then the inequality*

$$f(x) - f(x^0) \geqq (\mathrm{d}f)_{x^0} (x - x^0) = \langle \operatorname{grad} f(x^0), x - x^0 \rangle$$

holds for all $x \in C$.

Proof: First we will show that $\dfrac{f(x + s) - f(x)}{\|s\|}$ is monotonic. Consider two vectors s^1 and s^2 such that

(i) s^1 and s^2 have the same direction.
(ii) $0 < \|s^2\| \leqq \|s^1\|$.
(iii) $x + s^1 \in C$ and $x + s^2 \in C$, where x is any point in C.

(i) means $\dfrac{s^1}{\|s^1\|} = \dfrac{s^2}{\|s^2\|}$. If we put $\lambda = \dfrac{1}{\|s^1\|} (\|s^1\| - \|s^2\|)$, we have obviously $0 \leqq \lambda < 1$ and

$$x + s^2 = \lambda x + (1 - \lambda)(x + s^1).$$

Then we get, because of the convexity of f,

$$f(x + s^2) \leqq \lambda f(x) + (1 - \lambda) f(x + s^1).$$

Consequently,

$$f(x + s^2) - f(x) \leqq (1 - \lambda)\big(f(x + s^1) - f(x)\big) = \frac{\|s^2\|}{\|s^1\|}\big(f(x + s^1) - f(x)\big).$$

Dividing this inequality by $\|s^2\|$, we get

$$\frac{f(x + s^2) - f(x)}{\|s^2\|} \leqq \frac{f(x + s^1) - f(x)}{\|s^1\|}$$

which proves the monotonicity.

Let s be any vector with $\|s\| > 0$. If f is Fréchet differentiable at an interior point x^0, then

$$\frac{f(x^0 + s) - f(x^0)}{\|s\|} = (\mathrm{d}f)_{x^0}\left(\frac{s}{\|s\|}\right) + \frac{a(x^0, s)}{\|s\|}.$$

Now, in this equation, we calculate the limits in such a way that we do not change the direction of the vector s but the length of s tends to zero.

In this case the vector $\dfrac{s}{\|s\|}$ is constant, namely the unit vector in the direction of s. Thus we get

$$\lim_{\|s\| \to 0} \frac{f(x^0 + s) - f(x^0)}{\|s\|} = (\mathrm{d}f)_{x^0}\left(\frac{s}{\|s\|}\right).$$

From this follows, because of the monotocity of the difference quotient,

$$\frac{f(x^0 + s) - f(x^0)}{\|s\|} \geqq (\mathrm{d}f)_{x^0}\left(\frac{s}{\|s\|}\right).$$

We get the assertion of the theorem if we put $x = x^0 + s$. For the case $x = x^0$, the assertion is trivial. □

If the function f is concave and Fréchet differentiable on the open convex set C, then from Theorem 1.4.10 follows immediately that for any interior point $x^0 \in C$, we have

$$f(x) - f(x^0) \leqq (\mathrm{d}f)_{x^0} (x - x^0) = \langle \operatorname{grad} f(x^0), x - x^0 \rangle$$

for all $x \in C$.

Theorem 1.4.11. *Let a function f be Fréchet differentiable at all points of an open convex set C. If for any points $x, z \in C$ we have*

$$f(x) - f(z) \geqq (\mathrm{d}f)_z (x - z),$$

then f is a convex function.

Proof: We consider a real number λ, $0 \leqq \lambda \leqq 1$ and the point $z = \lambda x + (1 - \lambda) y$. Then z belongs to C if $x, y \in C$. Now, under the assumption of the theorem, we have

$$f(x) - f(z) \geqq (\mathrm{d}f)_z (x - z)$$

and

$$f(y) - f(z) \geqq (\mathrm{d}f)_z (y - z).$$

From this follows

$$\lambda f(x) + (1 - \lambda) f(y) \geqq f(z) + (\mathrm{d}f)_z \big(\lambda(x - z) + (1 - \lambda)(y - z)\big) = f(z).$$

Consequently, f is convex. □

At the end of this section, we will introduce the idea of a closed convex function. We have already proved that a convex function is continuous at any interior point of a convex set C, where the function is defined, but we have demonstrated with an example that this must not necessarily be so on the boundary. The following theorem gives a complete answer to this problem of continuity on the boundary.

Theorem 1.4.12. *Let f be a convex function defined on a convex set C. If b is a point on the boundary of C then we have*

$$\liminf_{x \to b} f(x) \leqq f(b).$$

Proof: If we take a point $x \in C$, then obviously $z = \frac{1}{2}(x + b) \in C$ if b is a point on the boundary. We can easily find $\lim_{x \to b} z = b$.

Because f is a convex function, we have

$$f(z) \leqq \frac{1}{2} f(x) + \frac{1}{2} f(b)$$

and from this follows

$$\liminf_{z \to b} f(z) = \liminf_{x \to b} f(x) \leqq \frac{1}{2} \liminf_{x \to b} f(x) + \frac{1}{2} f(b)$$

which is the assertion of the theorem. □

Theorem 1.4.12 may be used to complete a convex function in a certain sense. If we have a convex function f defined on a convex set C, we can complete the convex set C by adding those points x^0, for which $\liminf_{x \to x^0} f(x)$ is finite. In this way we construct a set

$C_0 \supseteqq C$. If we furthermore extend the function f on C_0 by defining the values by

$$f_0(x^0) = \liminf_{x \to x^0} f(x)$$

we get a function f_0 defined on C_0 which is an extension of f on C. We call the pair (f_0, C_0) the *closed hull* of (f, C).

Theorem 1.4.13. *If f is a convex function defined on a convex set C then f_0 and C_0 are convex.*

Proof: Let x^0 and y^0 be elements of C_0. Then, we have

$$f_0(x^0) = \liminf_{x \to x^0} f(x), \quad f_0(y^0) = \liminf_{y \to y^0} f(y).$$

Now, we consider two sequences $\{x^k\}$, $\{y^k\}$ where the elements of these sequences are interior points of C and $\lim_{k\to\infty} x^k = x^0$, $\lim_{k\to\infty} y^k = y^0$. If λ is a real number with $0 \leqq \lambda \leqq 1$, then $\{z^k\}$ with $z^k = \lambda x^k + (1 - \lambda)\, y^k$ is a sequence where its elements are also interior points of C and $\lim_{k\to\infty} z^k = z^0 = \lambda x^0 + (1 - \lambda)\, y^0$.

Obviously z^0 belongs to C_0 if $\liminf_{k\to\infty} f(z^k)$ is finite. If we use the construction of f_0, we get

$$\liminf_{k\to\infty} f(z^k) \leqq \lambda \liminf_{k\to\infty} f(x^k) + (1 - \lambda) \liminf_{k\to\infty} f(y^k) = \lambda f_0(x^0) + (1 - \lambda)\, f_0(y^0).$$

Consequently z^0 belongs to C_0 and thus C_0 is convex. Furthermore

$$f_0\big(\lambda x^0 + (1 - \lambda)\, y^0\big) = f_0(z^0) = \liminf_{z \to z^0} f(z) \leqq \lambda f_0(x^0) + (1 - \lambda)\, f_0(y^0),$$

showing that f_0 is a convex function defined on C_0. □

This result leads us to a definition which now follows.

Definition 1.4.3. We call a function f defined on a convex set C *closed*, if $f = f_0$ and $C = C_0$.

Even if a convex function is not closed, one can always construct the closed hull and from this point of view, we will in future always consider closed convex functions.

1.5. Conjugate of convex functions

In the construction of a theory of duality for convex programming, convex functions, the conjugate convex functions and their relations to convex functions and certain convex sets are of fundamental importance. For the representation of these relations, we refer to investigations of Fenchel [12] and to some generalizations by Karlin [20]. The construction of conjugate convex functions in this section will be done in such a way that we can easily generalize the results for topological vector spaces as you may find it in the work of Dieter [6].

Let f be a convex function defined on a convex set $C \subseteqq \mathbb{R}^n$. We consider the set

$$[f, C] = \{(\alpha, x) \in \mathbb{R} \times C : \alpha \geqq f(x)\}. \tag{1.5.1}$$

If f is a closed convex function then $[f, C]$ is a closed convex set contained in $\mathbb{R}^{n+1}$. In the sequel, we always consider closed convex functions. It is already known that through every boundary point of a closed convex set goes at least one tangent hyperplane. Using this property, it is possible to describe the set $[f, C]$ not only by it points (α, x) but also by all tangent hyperplanes as we have explained it by the end of Section 1.2. This is a reason why we define for $[f, C]$ a corresponding dual set $[f, C]^*$ which contains all such hyperplanes.

We now give the formal description of the set $[f, C]^*$.

Definition 1.5.1. The set of all nonvertical hyperplanes

$$H = \{(z, x) \in \mathbb{R} \times \mathbb{R}^n : \langle(-1, a), (z, x)\rangle - \gamma = 0\}$$

with the property

$$\langle(-1, a), (\alpha, x)\rangle - \gamma \leqq 0$$

for all $(\alpha, x) \in [f, C]$ is called the *corresponding conjugate set* of $[f, C]$. We denote this conjugate set by $[f, C]^*$.

As we have discussed in Section 1.2 every hyperplane H is uniquely defined by a pair $(\gamma, a) \in \mathbb{R} \times \mathbb{R}^n$. The direction of the vector $(-1, a)$ is orthogonal to H and γ defines the intercept made by the hyperplane on the vertical coordinate axis. Now we identify the pair (γ, a) with the hyperplane and from this point of view we understand $[f, C]^*$ as a subset of $\mathbb{R}^{n+1}$.

To help us understand this concept more clearly, let us construct an example.

We consider in $\mathbb{R}^1 = \mathbb{R}$ the closed convex set $C = [-2, 2] \subseteqq \mathbb{R}$ and a convex function $f(x) = x^2 + 2$ defined on C. Then the hyperplane

$$H = \{(z, x) \in \mathbb{R} \times \mathbb{R} : \langle(-1, 1), (z, x)\rangle + 1 = 0\}$$

defined by the pair $(\gamma, a) = (-1, 1)$ is an element of the conjugate set $[f, C]^*$. Thus, for any $(\alpha, x) \in [f, C]$, we have $\alpha \geqq x^2 + 2 \geqq -x + 1$. Consequently, $-\alpha - x + 1 = \langle(-1, -1), (\alpha, x)\rangle + 1 \leqq 0$ (see Fig. 1). A hyperplane which belongs to the conjugate set $[f, C]^*$ obviously lies completely under the convex set $[f, C]$. This is the case, if for all $(\alpha, x) \in [f, C]$, we have

$$\alpha \geqq \langle a, x\rangle - \gamma.$$

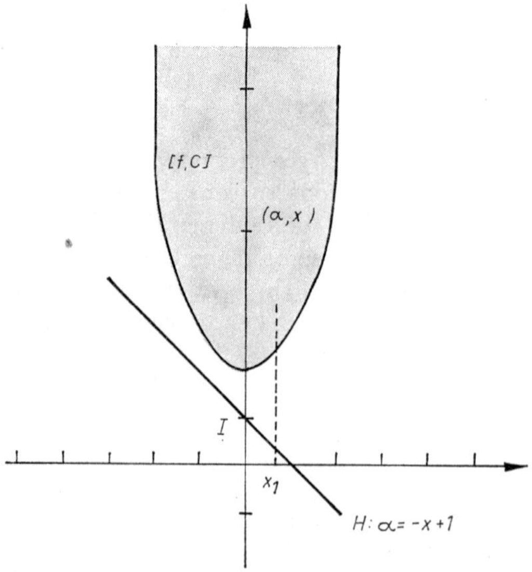

Fig. 1

Specially this inequality will be satisfied by points $(f(x), x) \in [f, C]$ with $x \in C$, i.e.

$$f(x) \geqq \langle a, x \rangle - \gamma, \quad x \in C.$$

Let $a \in \mathbb{R}^n$ be any fixed vector. Then $(-1, a)$ defines the direction of an infinite number of parallel hyperplanes. If we assume $(\gamma_1, a) \in [f, C]^*$, then obviously any hyperplane (γ, a) with $\gamma > \gamma_1$ also belongs to $[f, C]^*$. This means, for any $x \in C$, we have

$$\gamma > \gamma_1 \geqq \langle a, x \rangle - f(x)$$

which implies that $\langle a, x \rangle - f(x)$ is bounded from above over C. It follows that there exist a real number $\tilde{\gamma}$ with

$$\tilde{\gamma} = \sup_C \{\langle a, x \rangle - f(x)\}.$$

We can interpret $\tilde{\gamma}$ as the smallest number for which the hyperplane $(\tilde{\gamma}, a)$ with a fixed direction $(-1, a)$ belongs to the conjugate set $[f, C]^*$ i.e. besides $(\tilde{\gamma}, a)$ any hyperplane (γ, a) with $\gamma > \tilde{\gamma} = \sup_C \{\langle a, x \rangle - f(x)\}$ also belongs to $[f, C]^*$.

In our consideration, we started with a fixed vector $a \in \mathbb{R}^n$ with the property $(\gamma_1, a) \in [f, C]^*$ and then we construct the set of all hyperplanes with a direction $(-1, a)$ which also belong to $[f, C]^*$. Now we remove the restriction of a being fixed and define the set

$$C^* = \{a \in \mathbb{R}^n : \tilde{\gamma} = \sup_C \{\langle a, x \rangle - f(x)\} < \infty\}.$$

If $a \in C^*$ then obviously all hyperplanes (γ, a) with $\gamma \geqq \tilde{\gamma}$ belong to $[f, C]^*$. The set C^* contains all those direction vectors $(-1, a)$ of hyperplanes from $[f, C]^*$.

On C^* is now defined a function

$$f^*(a) = \tilde{\gamma} = \sup_C \{\langle a, x \rangle - f(x)\}.$$

If we use this notation we can represent the conjugate set by

$$[f, C]^* = \{(\sigma, a) \in \mathbb{R} \times C^* : \sigma \geqq f^*(a)\}. \tag{1.5.2}$$

We call f^* the *conjugate convex function of* f.

Lemma 1.5.1. *Let C be a closed convex set which contains interior points and f a closed convex function defined on C. Then C^* is a nonempty convex set and f^* a closed convex function with C^* as domain of definition.*

Proof: Because f is closed so also is the convex set $[f, C]$ closed. By Theorem 1.2.5, through every boundary point of $[f, C]$ goes a tangent hyperplane. Thus, if x^0 is an interior point of C, then there exist a tangent hyperplane on $[f, C]$ through the point $(f(x^0), x^0)$. If x^0 is an interior point then this hyperplane cannot be vertical and therefore it belongs to $[f, C]^*$ i.e. $C^* \neq \emptyset$.

Now we will prove the convexity of C^* and f^*. Consider the point $a = \lambda a^1 + (1-\lambda) a^2$ with $a^1, a^2 \in C^*$ and $0 \leqq \lambda \leqq 1$. We have

$$\sup_C \{\langle a, x \rangle - f(x)\} \leqq \lambda \sup_C \{\langle a^1, x \rangle - f(x)\} + (1-\lambda) \sup_C \{\langle a^2, x \rangle - f(x)\}.$$

Consequently $a \in C^*$ and C^* is convex. The above inequality also means

$$f^*(a) \leqq \lambda f^*(a^1) + (1-\lambda) f^*(a^2)$$

which shows that f^* is a convex function. Next, we have to show that f^* is a closed function. Let $\{a^k\}$ be a sequence of interior points of C^* which converge to a. Then we have for any $x \in C$

$$f^*(a^k) \geqq \langle a^k, x\rangle - f(x).$$

From this follows

$$\liminf_{k\to\infty} f^*(a^k) \geqq \liminf_{k\to\infty} \langle a^k, x\rangle - f(x) = \langle a, x\rangle - f(x) \quad \text{for any } x \in C.$$

Thus we have

$$\liminf_{k\to\infty} f^*(a^k) \geqq \sup_C \{\langle a, x\rangle - f(x)\} = f^*(a).$$

On the other hand from Theorem 1.4.12 and from the fact that a convex function is continuous at any interior point, we get

$$\liminf_{k\to\infty} f^*(a^k) \leqq f^*(a).$$

Altogether we have $\liminf_{k\to\infty} f^*(a^k) = f^*(a)$ implying that f^* is closed.

If f^* is a closed convex function defined on C^* then according to (1.5.1), the set

$$[f^*, C^*] = \{(\sigma, a) \in \mathbb{R} \times C^* : \sigma \geqq f^*(a)\}$$

is convex and closed. □

This result together with (1.5.2) implies that

$$[f, C]^* = [f^*, C^*].$$

Thus we arrive at a useful representation of the conjugate set $[f, C]^*$. Next we shall consider a relation between a convex function f and its conjugate convex function f^*.

Let us consider a boundary point $\big(f(x^0), x^0\big)$ of the convex set $[f, C]$ where x^0 is an interior point of C. By Theorem 1.2.5 there exists a tangent hyperplane on $[f, C]$ through $\big(f(x^0), x^0\big)$. This tangent hyperplane may be represented by the pair (γ, a^0) and it is obviously an element from the conjugate set $[f, C]^*$. With respect to this tangent hyperplane, we have

$$\langle a^0, x\rangle - f(x) \leqq \gamma \qquad \text{for } x \in C,$$

and

$$\langle a^0, x^0\rangle - f(x^0) = \gamma.$$

From these relations, there follows immediately

$$f^*(a^0) = \sup_C \{\langle a^0, x\rangle - f(x)\} = \langle a^0, x^0\rangle - f(x^0) = \gamma.$$

Consequently

$$f(x^0) + f^*(a^0) = \langle a^0, x^0\rangle. \tag{1.5.3}$$

Because $f^*(a^0) = \gamma$, we can say that the tangent hyperplane on $[f, C]$ through the boundary point $\big(f(x^0), x^0\big)$ is represented by a boundary point $\big(f^*(a^0), a^0\big)$ of the conjugate set $[f, C]^*$. The equation (1.5.3) gives us a further interpretation. Between the sets $[f, C]$ and $[f, C]^* = [f^*, C^*]$ there exists a symmetry which gives us the possibility to understand the point $\big(f(x^0), x^0\big)$ as a pair of parameters generating a hyperplane

$$H = \{(\sigma, a) \in \mathbb{R} \times \mathbb{R}^n : \langle(-1, x^0), (\sigma, a)\rangle - f(x^0) = 0\}.$$

The equation (1.5.3) means that the boundary point $(f^*(a^0), a^0)$ of $[f, C]^*$ belongs to the hyperplane H. If we now consider any point $(\sigma, a) \in [f, C]^*$, then, by (1.5.2), we have $\sigma \geqq f^*(a)$ and if we use the definition of f^* we have

$$\langle(-1, x^0), (\sigma, a)\rangle - f(x^0) \leqq \langle a, x^0\rangle - f^*(a) - f(x^0) \leqq 0\,.$$

Consequently $[f, C]^* \subseteqq H^+$ and thus H is a tangent hyperplane on $[f, C]^*$ through the boundary point $(f^*(a^0), a^0)$. Altogether we can say that every boundary point $(f^*(a^0), a^0)$ of the conjugate $[f, C]^*$ produces a tangent hyperplane on $[f, C]$ through a boundary point $(f(x^0), x^0)$ and conversely the point $(f(x^0), x^0)$ produces a tangent hyperplane on $[f, C]^*$ through the boundary point $(f^*(a^0), a^0)$. We call the pair (a^0, x^0) a convex conjugate pair of points. In view of (1.5.3), for any convex conjugate pair of points, we have

$$f(x^0) + f^*(a^0) = \langle a^0, x^0\rangle\,.$$

If we consider a tangent hyperplane $(f^*(a^0), a^0) \in [f, C]^*$ on the set $[f, C]$ it may go through more than one boundary point of $[f, C]$. We denote by $B(a^0)$, the set of all boundary points of $[f, C]$ which belong to the hyperplane $(f^*(a^0), a^0)$. Then every element $(f(x), x) \in B(a^0)$ produces a tangent hyperplane on $[f, C]^*$ through $(f^*(a^0), a^0)$.

By a simple example we shall now demonstrate this idea.

Let us consider the interval $C = [1, 2]$ and the convex function $f(x) = x - 1$ defined on C. The representation of $[f, C]$ is given in Fig. 2.

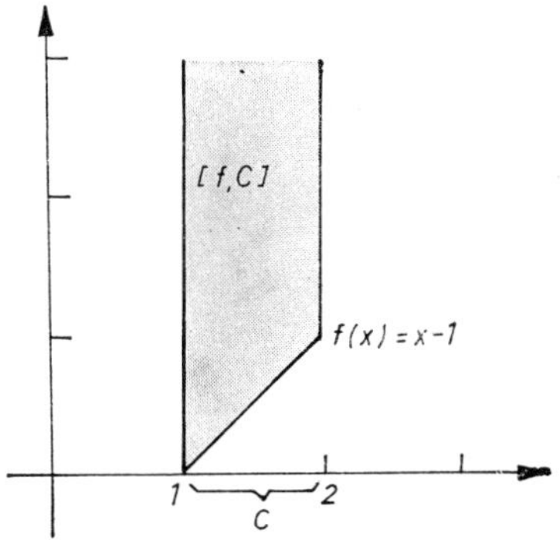

Fig. 2

Further more it is obvious that $C^* = \mathbb{R}$ and

$$f^*(a) = \sup_{x\in[1,2]} \{(a - 1)\, x + 1\} = \begin{cases} 2a - 1 & \text{if } a \geqq 1\,, \\ a & \text{if } a \leqq 1\,. \end{cases}$$

The representation of the conjugate set $[f, C]^* = [f^*, C^*]$ is given in Fig. 3 If we choose $a^0 = 1$, then the tangent hyperplane $(f^*(a^0), a^0) = (1, 1)$ on $[f, C]$ goes through at least one point $(f(x^0), x^0)$. This means

$$\langle(-1, 1), (f(x^0), x^0)\rangle - 1 = -f(x^0) + x^0 - 1 = 0\,.$$

Since every point $x \in C$ satisfies this equation, we have

$$B(a^0) = B(1) = \{(f(x), x) : x \in [1, 2]\}$$

and the tangent hyperplane on $[f, C]$ is defined by $\alpha = x - 1$. On the other hand every point $(f(x), x) \in B(1)$ produces a tangent hyperplane on $[f^*, C^*]$ through $(f^*(a^0), a^0) = (1, 1)$; namely

$$\langle(-1, x), (f^*(a^0), a^0)\rangle - f(x) = a^0 x - f^*(a^0) - x + 1 = 0\,.$$

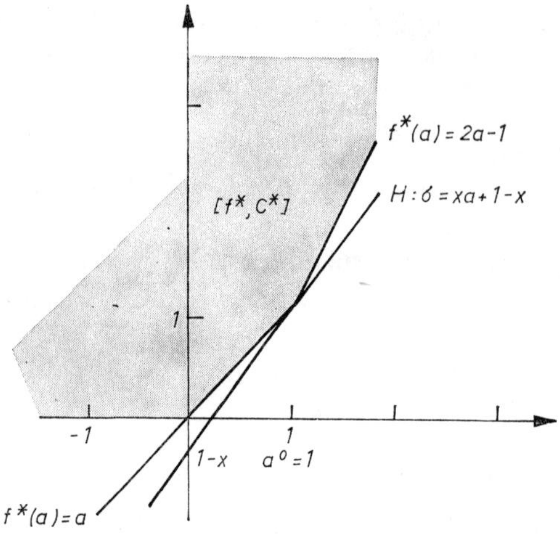

Fig. 3

Using this equation, we can calculate a^0 as follows:

$$a^0(x-2) = x-2 \qquad \text{if } a^0 \geqq 1,$$

$$a^0(x-1) = x-1 \qquad \text{if } a^0 \leqq 1$$

and in consequence $a^0 = 1$.

The tangent hyperplane through $\big(f^*(1), 1\big) = (1, 1)$ has the form

$$\langle(-1, x), (\sigma, a)\rangle - f(x) = -\sigma + ax + 1 - x = 0$$

or

$$\sigma = \sigma(a) = xa + 1 - x.$$

This is a family of tangent hyperplanes on $[f^*, C^*]$ through the boundary point $(1, 1)$ for each $x \in [1, 2]$. In case $x = 1$, we get $\sigma = a$ and the tangent hyperplane contains all points $\big(f^*(a), a\big) \in [f^*, C^*]$ if $a \leqq 1$. In case $x = 2$, we get $\sigma = 2a - 1$ and the tangent hyperplane on $[f^*, C^*]$ contains all points $\big(f^*(a), a\big)$ if $a \geqq 1$. The rest of the hyperplanes for $1 < x < 2$ contains exactly one boundary point from $[f^*, C^*]$, namely $(1, 1)$. Now, we shall illustrate with an example under which conditions the correspondence of tangent hyperplanes $\big(f^*(a^0), a^0\big)$ and $\big(f(x^0), x^0\big)$ produced by a convex conjugate pair of points (a^0, x^0) is uniquely reversible.

Consider $C = [1, \infty)$ and the convex function $f(x) = \dfrac{1}{x}$ defined on C. The conjugate convex function f^* has the form

$$f^*(a) = \begin{cases} -2\sqrt{-a} & \text{if } 0 \geqq a \geqq -1, \\ a - 1 & \text{if } -1 \geqq a, \end{cases}$$

where f^* is defined on $C^* = (-\infty, 0]$.

If we choose $a^0 = -\dfrac{1}{4}$, then we get a tangent hyperplane

$$\left(f^*\left(-\frac{1}{4}\right), -\frac{1}{4}\right) = \left(-1, -\frac{1}{4}\right) \in [f^*, C^*] \quad \text{on} \quad [f, C]$$

through a boundary point $(f(x^0), x^0)$, which satisfies the equation

$$\left\langle\left(-1, -\frac{1}{4}\right), \left(f(x^0), x^0\right)\right\rangle + 1 = -f(x^0) - \frac{1}{4}x^0 + 1 = -\frac{1}{x^0} - \frac{1}{4}x^0 + 1 = 0.$$

For this equation $x^0 = 2$ is the uniquely defined solution and the tangent hyperplane $\alpha = -\frac{1}{4}x + 1$ on $[f, C]$ goes through $\left(f(2), 2\right) = \left(\frac{1}{2}, 2\right)$ which is the only boundary point of $[f, C]$ contained in the tangent hyperplane (see Fig. 4).

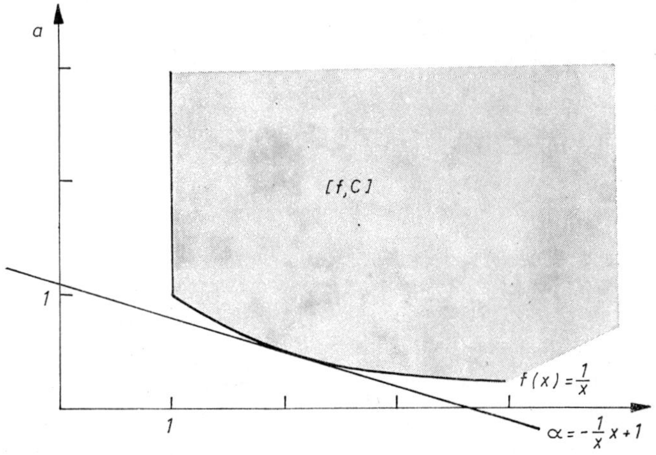

Fig. 4

On the other hand, the point $\left(\frac{1}{2}, 2\right) \in [f, C]$ produces a tangent hyperplane on $[f^*, C^*]$ through a point $\left(f^*(a^0), a^0\right)$. We can calculate a^0 from the equation

$$\left\langle(-1, 2), \left(f^*(a^0), a^0\right)\right\rangle - \frac{1}{2} = -f^*(a^0) + 2a^0 - \frac{1}{2} = 0.$$

In case $a^0 \leqq -1$, the equation

$$f(a^0) = a^0 - 1 = 2a^0 - \frac{1}{2}$$

is not solvable. Consequently $a^0 \geqq -1$. Then it follows from

$$f(a^0) = -2\sqrt{-a^0} = 2a^0 - \frac{1}{2}$$

that $a^0 = -\frac{1}{4}$ and a^0 is uniquely defined.

Thus, the tangent hyperplane $\sigma = 2a - \frac{1}{2}$ on $[f^*, C^*]$ goes through exactly one point; namely $\left(f^*\left(-\frac{1}{4}\right), -\frac{1}{4}\right) = \left(-1, -\frac{1}{4}\right)$ (see Fig. 5). These two examples lead us to the following theorem.

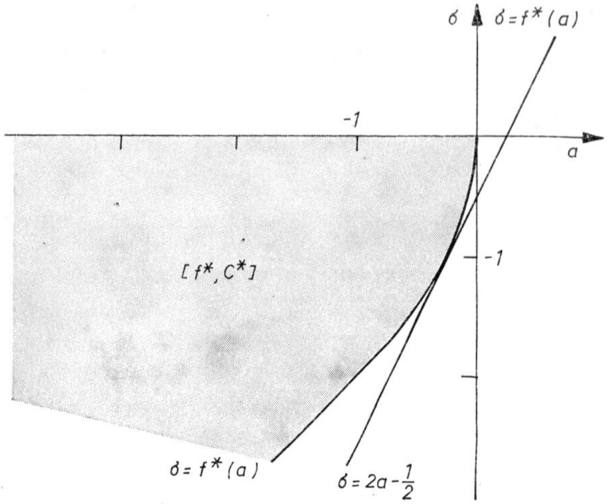

Fig. 5

Theorem 1.5.1. *Let (a^0, x^0) be a pair of convex conjugate points with respect to the convex function f and its conjugate f^*. If $a^0 \in C^*$ and $x^0 \in C$ are interior points and if there exist neighborhoods $S_\varepsilon(a^0) \subseteqq C^*$ and $S_\varepsilon(x^0) \subseteqq C$ such that f and f^* are strictly convex in those neighborhoods, then the correspondence between a^0 and x^0 is unique and the tangent hyperplanes on $[f, C]$ through $\big(f(x^0), x^0\big)$ and the corresponding tangent hyperplanes on $[f^*, C^*]$ through $\big(f^*(a^0), a^0\big)$ are uniquely defined.*

Proof: We consider the hyperplane $\big(f^*(a^0), a^0\big)$ and assume that this hyperplane goes through $\big(f(x^0), x^0\big)$ and $\big(f(x^1), x^1\big)$ with $x^0 < x^1$. Because of the convexity of $[f, C]$, for any λ with $0 \leqq \lambda \leqq 1$, the points

$$\big(f(x), x\big) = \lambda\big(f(x^0), x^0\big) + (1 - \lambda)\,\big(f(x^1), x^1\big)$$

are contained in $[f, C]$. Furthermore, together with $\big(f(x^0), x^0\big)$ and $\big(f(x^1), x^1\big)$ all the points $\big(f(x), x\big)$ belong to the tangent hyperplane $\big(f^*(a^0), a^0\big)$ and thus the points $\big(f(x), x\big)$ are boundary points of $[f, C]$. Consequently, the function f is linear in $x^0 \leqq x \leqq x^1$.

This contradicts the assumption that there exists a sphere $S_\varepsilon(x^0) \subseteqq C$ such that f is strictly convex in $S_\varepsilon(x^0)$.

The assertion of the theorem being symmetric, the theorem is proved. □

From the relation (1.5.3) for a convex conjugate pair of points (a^0, x^0), we get immediately a further result:

If a^0 and x^0 are interior points of C^* and C respectively and if the functions f^* and f are Fréchet-differentiable at a^0 and x^0 respectively, then it follows immediately from (1.5.3) that

$$(\mathrm{d}f^*)_{a^0} = \operatorname{grad} f^*(a^0) = x^0,$$

$$(\mathrm{d}f)_{x^0} = \operatorname{grad} f(x^0) = a^0.$$

This means under the given conditions, we can get a convex conjugate pair of points by calculating Fréchet derivatives of f^* and f. In our second example, we have

$$(\mathrm{d}f^*)_{a=-\frac{1}{4}} = \mathrm{d}\big(-2\sqrt{-a}\big)_{a=-\frac{1}{4}} = 2 = x^0,$$

$$(\mathrm{d}f)_{x=2} = \mathrm{d}\left(\frac{1}{x}\right)_{x=2} = -\frac{1}{4} = a^0.$$

In the theory of duality of mathematical programming, the concept of duality as we usually understand is that we start with a primary problem and construct a dual problem by applying a suitable transformation on the primary problem. The principle of duality is also characterized in such a way that by applying the transformation twice, we reproduce the primary problem. If we take, for such a transformation, the transition from a convex function f to its conjugate f^*, we get the following lemma.

Lemma 1.5.2. *If f is a closed convex function defined on a closed convex set C, then we have*

$$[f, C]^{**} = [f^*, C^*]^* = [f, C].$$

Proof: First we will show that $[f, C]$ is contained in $[f^*, C^*]^*$. Consider any element $(\alpha, x) \in [f, C]$.

Using the definition of f^*, we have for all $a \in C^*$

$$f^*(a) \geqq \langle a, x\rangle - f(x).$$

Since $\alpha \geqq f(x)$, this inequality implies

$$\alpha \geqq \langle a, x\rangle - f^*(a) \tag{1.5.4}$$

for any $a \in C^*$.

This means the supremum of the right-hand side of (1.5.4) over C^* exists and we have

$$\alpha \geqq \sup_{C^*} \{\langle a, x\rangle - f^*(a)\} = (f^*)^*(x). \tag{1.5.5}$$

Then we can represent the conjugate set of $[f^*, C^*]$ in the form

$$[f^*, C^*]^* = \{(\alpha, x) \in \mathbb{R} \times C^{**}: \alpha \geqq (f^*)^*(x)\}$$

with

$$C^{**} = \{x \in \mathbb{R}^n: \sup_{C^*} \{\langle a, x\rangle - f^*(a)\} < \infty\}.$$

Now it follows from (1.5.4) that any element $x \in C$ also belongs to C^{**} and if $(\alpha, x) \in [f, C]$, it follows from (1.5.5) that $(\alpha, x) \in [f^*, C^*]^*$. Thus

$$C \subseteqq C^{**} \quad \text{and} \quad [f, C] \subseteqq [f^*, C^*]^*.$$

To complete the proof of this theorem we will show that if (α_0, x^0) is not an element in $[f, C]$, then this element cannot belong to $[f^*, C^*]^*$.

If $(\alpha_0, x^0) \notin [f, C]$ the point (α_0, x^0) has a positive distance from $[f, C]$ because the set $[f, C]$ is closed. This means $\delta([f, C], \{(\alpha_0, x^0)\}) > 0$. By the separation Theorem 1.2.2, there exists a hyperplane (γ, a) which separated $[f, C]$ from (α_0, x^0) strictly.

With respect to this hyperplane, we have

$$\langle(-1, a), (\alpha, x)\rangle - \gamma < 0 \quad \text{for any } (\alpha, x) \in [f, C] \tag{1.5.6}$$

and

$$\langle(-1, a), (\alpha_0, x^0)\rangle - \gamma > 0. \tag{1.5.7}$$

Especially from (1.5.6), it follows that for any $x \in C$

$$f(x) > \langle a, x\rangle - \gamma$$

and consequently

$$\gamma \geqq \sup_{C} \{\langle a, x\rangle - f(x)\} = f^*(a).$$

This means the hyperplane (γ, a) belongs to $[f^*, C^*]$.

On the other hand it follows from $\gamma \geqq f^*(a)$ and (1.5.7) that

$$\alpha_0 < \langle a, x^0\rangle - \gamma \leqq \langle a, x^0\rangle - f^*(a).$$

Consequently, it is impossible that (α_0, x^0) is an element in $[f^*, C^*]^*$ and this proves the theorem. □

The Lemma 1.5.2 shows that $C^{**} = C$ and that $f(x)$ is the conjugate convex function of $f^*(a)$. We have

$$(f^*)^*(x) = f^{**}(x) = f(x) = \sup_{C^*} \{\langle a, x\rangle - f^*(a)\}.$$

The assumption in Lemma 1.5.2 that f has to be a closed convex function is essential. Let us consider for example a function $f(x) = \sqrt{\|x\|}$ defined on $C = \mathbb{R}^n$. Then obviously

$$f^*(0) = \sup_{\mathbb{R}^n} \left\{\langle 0, x\rangle - \sqrt{\|x\|}\right\} = 0.$$

If we choose $a \neq 0$, following the Schwarz inequality $|\langle a, x\rangle| \leqq \|a\| \, \|x\|$ we find such an element $x \in \mathbb{R}^n$ that $\|a\| \geqq |\langle a, x\rangle| \geqq \frac{\|a\|}{2}$ and $\|x\| = 1$. Moreover one can easily see according to the definition of conjugate functions that $f^*(a) + f(x) \geqq \langle a, x\rangle$. Then it follows for any $t > 0$

$$f^*(a) + f(tx) = f^*(a) + \sqrt{t} \geqq t\langle a, x\rangle > t\,\frac{\|a\|}{2}.$$

Thus $f^*(a) > \sqrt{t}\left(\sqrt{t}\,\frac{\|a\|}{2} - 1\right) \xrightarrow[t\to\infty]{} \infty$. Altogether we have

$$f^*(a) = \delta_{\{0\}}(a) = \begin{cases} 0, & \text{if } a = 0, \\ \infty, & \text{if } a \neq 0 \end{cases}$$

and $C^* = \mathbb{R}^n$. On the other hand it is obvious that

$$f^{**}(x) = \sup_{C^*} \{\langle a, x\rangle - \delta_{\{0\}}(a)\} \equiv 0$$

thus $f^{**} \neq f$. This result depends on the fact, that $f(x) = \sqrt{\|x\|}$ is not a convex function.

In the study of convex programming it is necessary to make use of the behaviour of a sum of two convex functions when transformed to their conjugates.

Let f_1 and f_2 be closed convex functions defined on convex sets C_1 and C_2. Then $[f_1, C_1]$ and $[f_2, C_2]$ are closed convex sets where $[f_1^*, C_1^*]$ and $[f_2^*, C_2^*]$ are the corresponding conjugate sets.

We define

$$\begin{aligned} T &= [f_1^*, C_1^*] + [f_2^*, C_2^*] \\ &= \{(\gamma, a) \in \mathbb{R}^{n+1} : \gamma = \gamma_1 + \gamma_2;\ a = a^1 + a^2;\ (\gamma_i, a^i) \in [f_i^*, C_i^*];\ i = 1, 2\}. \end{aligned}$$

Let $\overline{T}$ denote the closure of T.

Lemma 1.5.3. *Let f_1 and f_2 be closed convex functions defined on C_1 and C_2. If $C = C_1 \cap C_2 \neq \emptyset$, then for the convex function $f(x) = f_1(x) + f_2(x)$ defined on C,*

$$[f, C]^* = \overline{T} = [f^*, C^*]$$

with

$$C_1^* + C_2^* \subseteqq C^* \subseteqq \overline{C_1^* + C_2^*}$$

and

$$f^*(a) = \inf \{f_1^*(a^1) + f_2^*(a^2) \colon a = a^1 + a^2;\ a^i \in C_i^*,\ i = 1, 2\}.$$

This lemma says that for any element $a = a^1 + a^2 \in C_1^* + C_2^*$, we always have

$$f^*(a) \leqq f_1^*(a^1) + f_2^*(a^2).$$

Proof: The proof of this lemma consists of three steps:

(i) $\overline{T} \subseteqq [f, C]^*$,

(ii) $[f, C]^* = \overline{T}$,

(iii) $\overline{T} = [f^*, C^*]$.

To prove (i), we choose any element

$$(\gamma, a) = (\gamma_1 + \gamma_2, a^1 + a^2) \in T.$$

Then we have for all $x \in C_1 \cap C_2$,

$$\gamma_1 \geqq f_1^*(a^1) \geqq \langle a^1, x\rangle - f_1(x)$$

$$\gamma_2 \geqq f_2^*(a^2) \geqq \langle a^2, x\rangle - f_2(x).$$

Adding these two inequalities

$$\gamma = \gamma_1 + \gamma_2 \geqq \langle a^1 + a^2, x\rangle - f_1(x) - f_2(x) = \langle a, x\rangle - f(x)$$

for all $x \in C = C_1 \cap C_2$. Consequently

$$\gamma \geqq \sup_{C_1 \cap C_2} \{\langle a, x\rangle - f(x)\} = f^*(a).$$

Thus the element $(\gamma, a) = (\gamma_1 + \gamma_2, a^1 + a^2)$ belongs to $[f, C]^*$ and $T \subseteqq [f, C]^*$. Because the function $f = f_1 + f_2$ is closed it follows that the convex set $[f, C]^*$ is closed and $\overline{T} \subseteqq [f, C]^*$.

To prove (ii), we consider conjugate sets and show first $T^* \subseteqq [f, C]^{**}$. We notice that we have $[f, C]^{**} = [f, C]$ by Lemma 1.5.2.

By Definition 1.5.1, an element (α, x) belongs to T^* if and only if for any $(\gamma, a) = (\gamma_1 + \gamma_2, a^1 + a^2) \in T$, we have

$$0 \geqq \langle(-1, x), (\gamma, a)\rangle - \alpha = \langle a^1 + a^2, x\rangle - \gamma_1 - \gamma_2 - \alpha.$$

This inequality is obviously equivalent to

$$\alpha \geqq \langle a^1, x\rangle - f_1^*(a^1) + \langle a^2, x\rangle - f_2^*(a^2)$$

for any $a^1 \in C_1^*$ and any $a^2 \in C_2^*$.

Consequently we have

$$\alpha \geqq \sup_{C_1^*} \{\langle a^1, x\rangle - f_1^*(a^1)\} + \sup_{C_2^*} \{\langle a^2, x\rangle - f_2^*(a^2)\}.$$

Because $f_i^{**} = f_i$, $i = 1, 2$, it follows that

$$\alpha \geqq f_1(x) + f_2(x) = f(x)$$

where α can be an upper bound only if $x \in C_1 \cap C_2 = C$. Thus the element $(\alpha, x) \in T^*$ belongs to $[f, C]$ i.e.

$$T^* \subseteqq [f, C]. \tag{1.5.8}$$

By the definition of conjugate sets (1.5.8) implies

$$[f, C]^* \subseteqq T^{**}. \tag{1.5.9}$$

Now an element (γ, a) belongs to T^{**} if and only if for any $(\alpha, x) \in T^*$, we have

$$0 \geqq \langle(-1, a), (\alpha, x)\rangle - \gamma = -\alpha + \langle a, x\rangle - \gamma$$

i.e.

$$\gamma \geqq \langle a, x\rangle - \alpha.$$

This inequality is equivalent to

$$\gamma \geqq \langle a, x\rangle - f(x) = \langle a, x\rangle - f_1(x) - f_2(x)$$

for any $x \in C = C_1 \cap C_2$. From this follows

$$\gamma \geqq \sup_C \{\langle a, x\rangle - f(x)\} = f^*(a).$$

Thus the element $(\gamma, a) \in T^{**}$ belongs to $[f, C]^*$ i.e. $T^{**} \subseteqq [f, C]^*$. This result together with (1.5.9) implies $T^{**} = [f, C]^*$ where T^{**} is closed because $[f, C]^*$ is closed.

Furthermore, since $\overline{T}$ is a closed convex set, $\overline{T}^{**} = \overline{T}$. Using (i), we have $T \subseteqq \overline{T} \subseteqq [f, C]^* = T^{**}$.

If we apply on this enclosing relation the star transformation twice, we get

$$T^{**} \subseteqq \overline{T}^{**} = \overline{T} \subseteqq [f, C]^{***} = [f, C]^* = T^{**}.$$

Consequently, $\overline{T} = T^{**} = [f, C]^*$ and this proves (ii).

To prove (iii), we consider f_1^* and f_2^*. If these two conjugate functions are closed then obviously f^* is closed and thus $[f^*, C^*]$ is a closed convex set.

Let (γ, a) be an element in T. Then

$$\begin{aligned}\gamma = \gamma_1 + \gamma_2 &\geqq f_1^*(a^1) + f_2^*(a^2)\\ &\geqq \inf\{f_1^*(a^1) + f_2^*(a^2) : a = a^1 + a^2; a^i \in C_i^*, i = 1, 2\} = f^*(a).\end{aligned}$$

This means $T \subseteqq [f^*, C^*]$ and because $[f^*, C^*]$ is a closed set we also have $\overline{T} \subseteqq [f^*, C^*]$.

To show the reverse relation, we consider an element $(\gamma, a) \in [f^*, C^*]$ for which $\gamma \geqq f^*(a)$. Now we assume that (γ, a) does not belong to T and consider any representation of $a = a^1 + a^2$ with $a^i \in C_i^*$, $i = 1, 2$.

If we put $\gamma_1 = f_1^*(a^1)$, $\gamma_2 = \gamma - f_1^*(a^1)$ then we have $\gamma_1 + \gamma_2 = \gamma$.

Since (γ, a) is not an element in T

$$\gamma_2 = \gamma - f_1^*(a^2) < f_2^*(a^2).$$

From this follows

$$\gamma \leqq \inf\{f_1^*(a^1) + f_2^*(a^2) : a = a^1 + a^2; a^i \in C_i^*, i = 1, 2\} = f^*(a).$$

Thus $(\gamma, a) \in [f^*, C^*]$ implies $\gamma \geqq f^*(a)$ and $(\gamma, a) \notin T$ implies $\gamma \leqq f^*(a)$. Consequently all elements $(\gamma, a) \in [f^*, C^*]$, which do not belong to T have the form $(\gamma, a) = (f^*(a), a)$. This means that these elements are boundary points of $[f^*, C^*]$.

Thus $(\gamma, a) \in \overline{T}$ leading to $[f^*, C^*] \subseteqq \overline{T}$. Finally we have $[f^*, C^*] = \overline{T}$ which proves the lemma. □

For the theory of convex programming we usually make our discussions in the framework of concave functions. Let D be a convex set in $\mathbb{R}^n$ and g a concave function defined on D. We call g closed if the convex function $-g$ is closed. Then the set

$$[g, D] = \{(\alpha, x) \in \mathbb{R} \times D: \alpha \leqq g(x)\}$$

will also be closed.

Definition 1.5.2. The set of all nonvertical hyperplanes

$$H = \{(z, x) \in \mathbb{R} \times \mathbb{R}^n: \langle(-1, a), (z, x)\rangle - \gamma = 0\}$$

with the property

$$\langle(-1, a), (\alpha, x)\rangle - \gamma \geqq 0$$

for any $(\alpha, x) \in [g, D]$ is called the *conjugate set* to $[g, D]$. This conjugate set we denote by $[g, D]^*$.

From the geometrical point of view all those hyperplanes which belong to $[g, D]^*$ lie completely above the convex set $[g, D]$. Thus an element (γ, a) belongs to $[g, D]^*$ if and only if for any $x \in D$, we have

$$\gamma \leqq \langle a, x\rangle - g(x).$$

From this follows

$$\gamma \leqq \inf_D \{\langle a, x\rangle - g(x)\} = \tilde{\gamma}$$

and $\tilde{\gamma}$ is the greatest number such that $(\tilde{\gamma}, a)$ belongs to $[g, D]^*$. For any real number γ with $\gamma \leqq \tilde{\gamma}$, the element (γ, a) belongs to $[g, D]^*$.

We now define

$$D^* = \{a \in \mathbb{R}^n: \inf_D \{\langle a, x\rangle - g(x)\} > -\infty\}.$$

With respect to D^*, we define the conjugate concave function

$$g^*(a) = \inf_D \{\langle a, x\rangle - g(x)\}, \qquad a \in D^*.$$

D^* is a convex set and together with g the function g^* is concave and closed.

For $[g, D]^*$, we get the representation

$$[g, D]^* = \{(\gamma, a) \in \mathbb{R} \times D^*: \gamma \leqq g^*(a)\} = [g^*, D^*].$$

Furthermore, we remark that for concave functions as well, the results we have obtained above for convex functions hold; namely

$$[g, D]^{**} = [g^*, D^*]^* = [g, D]$$

and

$$D^{**} = D, \, g^{**}(x) = \inf_{D^*} \{\langle a, x\rangle - g^*(a)\} = g(x).$$

All results we have obtained above concerning convex and conjugate convex functions as well as concave functions hold in case we generalize our considerations and replace the finite dimensional space $\mathbb{R}^n$ by any linear Banach space E. Relating to this we may refer to Moreau [27], [28], [29]. If f is a real valued convex function defined on a convex set $C \subseteqq E$ we have to construct the conjugate convex function f^* on $C^* \subseteqq E^*$, where E, E^* are dual paired spaces i.e. E^* contains all continuous linear functionals defined on E. Let $x^* \in E$ be any linear functional. If we denote $x^*(x) = \langle x^*, x\rangle$ then, we have

$$f^*(x^*) = \sup_C \{\langle x^*, x\rangle - f(x)\}$$

and

$$C^* = \left\{x^* \in E^*: \sup_C \{\langle x^*, x\rangle - f(x)\} < \infty\right\}.$$

Obviously the following relations holds:

(i) $f(x) + f^*(x^*) \geqq \langle x^*, x\rangle$ for all (x, x^*) $C \times C^*$.

(ii) $(f + \alpha)^* (x^*) = \sup\limits_C \{\langle x^*, x\rangle - f(x) - \alpha\} = f^*(x^*) - \alpha$ for any $\alpha \in \mathbb{R}$.

(iii) $(\alpha f^*) (x^*) = \sup\limits_C \{\langle x^*, x\rangle - \alpha f(x)\} = \alpha f^* \left(\dfrac{x^*}{\alpha}\right)$ for any $\alpha > 0$.

(iv) If $f_1(x) = f\left(\dfrac{x}{\alpha}\right)$, $\alpha > 0$ we find

$$f_1^*(x^*) = \sup_C \left\{\langle x^*, x\rangle - f\left(\frac{x}{\alpha}\right)\right\} = f^*(\alpha x^*).$$

(v) If $f_1(x) = f(x) - \langle x_0^*, x\rangle$ we find

$$f_1^*(x^*) = \sup_C \{\langle x^* + x_0^*, x\rangle - f(x)\} = f^*(x^* + x_0^*).$$

Especially from (i), it follows that for any $x \in C$

$$f^{**}(x) \leqq f(x)$$

and moreover the Lemma 1.5.2 still holds.

2. Theory of duality

2.1. Fundamental theorems

In view of constructing a general theory of duality of mathematical optimization, we consider what we call a Lagrangian form which produced two corresponding dual problems. Then we will attempt to find relations between the Lagrangian form and the dual pair of problems. Such relations are expressed in the form of duality theorems, saddle-point theorems and minimax theorems. We will confine our attention to finite dimensional Euclidean vector spaces and refer to a paper by Vogel [33]. The situation will be such that we always assume the existence of solutions of one of the three problems, duality problem, saddle-point problem and minimax problem, and try to find a possible equivalence to one another. If we specially consider a concrete Lagrangian form to construct a special dual pair of problems, then of course the existence of solutions of at least one of the three problems is an important requirement. Because if we know the existence of solutions of one problem, then we will know, through the equivalence relations, the existence of solutions of other problems. Inquiries in this direction will be made in Chapter 3.

Let $\mu(x)$ and $\nu(y)$ be real valued functions defined respectively on $M \subseteqq \mathbb{R}^n$ and $N \subseteqq \mathbb{R}^m$. If for any $(x, y) \in M \times N$, we have

$$\mu(x) \leqq \nu(y),$$

then a supremum of μ and an infimum of ν exist and

$$\sup_M \mu(x) \leqq \inf_N \nu(y).$$

In the theory of mathematical programming, the following results are of high theoretical value:

(A$_1$) Under certain necessary and sufficient conditions we have

$$\sup_M \mu(x) = \inf_N \nu(y). \tag{2.1.1}$$

A statement of type (A$_1$) is called a *duality theorem.*

(A$_2$) Let $\psi(x, y)$ be a real valued function defined on $K \times L \subseteqq \mathbb{R}^n \times \mathbb{R}^m$. Then under certain necessary and sufficient conditions we have

$$\sup_K \inf_L \psi(x, y) = \inf_L \sup_K \psi(x, y). \tag{2.1.2}$$

A statement of type (A$_2$) is called a *minimax theorem.*

(A_3) Under certain conditions a pair of vectors $(x^0, y^0) \in K \times L$ exists such that for any $(x, y) \in K \times L$, we have

$$\varphi(x, y^0) \leqq \varphi(x^0, y^0) \leqq \varphi(x^0, y). \tag{2.1.3}$$

A statement of type (A_3) is called a *saddle-point theorem* and the pair (x^0, y^0) is called a *saddle point* of φ with the saddle-point coordinates x^0 and y^0.

The main purpose of the theory of duality in mathematical programming is to make inquiries about a corresponding pair of optimization problems, namely the primal problem to find

$$W = \sup_M \mu(x) \tag{P}$$

and the dual problem to find

$$W^* = \inf_N \nu(y) \tag{D}$$

and to discover relations between corresponding duality, minimax and saddle-point theorems.

Suppose for the problems (P) and (D), there exists a pair $(x^0, y^0) \in M \times N$ such that

$$\sup_M \mu(x) = \max_M \mu(x) = \mu(x^0),$$

$$\inf_N \nu(y) = \min_N \nu(y) = \nu(y^0).$$

Then we say that the problems (P) and (D) are realizable with respect to M and N respectively. In this case, we call x^0 and y^0 *optimal solutions*. To recognize this idea, we can augment the problems (A_1) to (A_3) by two further problems.

(A_4) Find necessary and sufficient conditions such that the problems (P) and (D) are realizable.
Assertions in this direction are called *realization theorems*.

(A_5) Construct methods to find optimal solutions of realizable problems, i.e. to define an optimal program by which we mean a set of instructions for actions with a view to solving a practical problem.

While the problems (A_1) to (A_4) are of qualitative nature, the problems (A_5) is a quantitative one which is important for solving practical concrete problems. In this monograph we are mainly concerned with qualitative problems.

2.2. Relations between saddle-point and minimax theorem

In this section we confine to the problem of finding relations between duality theorem (A_1), minimax theorem (A_2) and saddle-point theorem (A_3). For this purpose, there is a natural way to explore. If $\varphi(x, y)$ is a real valued function defined on $K \times L \subseteqq \mathbb{R}^n \times \mathbb{R}^m$, then we define

$$\mu(x) = \inf_L \varphi(x, y),$$

$$\nu(v) = \sup_K \varphi(x, y).$$

Let $M \times N \subseteqq K \times L$ be the domain of definition of the function μ and ν. This means M contains all points $x \in K$, for which the infimum exists and N contains all points $y \in L$

for which the supremum exists. The sets M and N may be empty or we may have the other extreme case $M = K$ and $N = L$. In a general case M and N will be properly contained in K and L.

Obviously we have, for any $(x, y) \in M \times L$,

$$\mu(x) \leqq \psi(x, y)$$

and for any $(x, y) \in K \times N$

$$\psi(x, y) \leqq \nu(y).$$

From this follows for any $(x, y) \in M \times N$

$$\mu(x) \leqq \psi(x, y) \leqq \nu(y). \tag{2.2.1}$$

Consequently there exists a supremum for $\mu(x)$ over M and an infimum for $\nu(y)$ over N. Thus we are able to formulate a primal problem (P) for the function μ and a dual problem (D) for the function ν. If these two problems are realizable, i.e. if there exists a pair $(x^0, y^0) \in M \times N$ such that

$$\sup_M \mu(x) = \max_M \mu(x) = \mu(x^0),$$

$$\inf_N \nu(y) = \min_N \nu(y) = \nu(y^0)$$

and if

$$\inf_L \psi(x, y) = \min_L \psi(x, y),$$

$$\sup_K \psi(x, y) = \max_K \psi(x, y)$$

then theorem of duality,

$$\mu(x^0) = \max_M \mu(x) = \min_N \nu(y) = \nu(y^0)$$

is obviously equivalent to minimax theorem,

$$\max_M \min_L \psi(x, y) = \min_N \max_K \psi(x, y).$$

In general, we find the following connection between minimax theorem and saddle-point theorem.

Theorem 2.2.1. *If there exists a minimax point $(x^0, y^0) \in K \times L$ such that*

$$\psi(x^0, y^0) = \max_K \min_L \psi(x, y) = \min_L \max_K \psi(x, y)$$

then for $\psi(x, y)$, a saddle point exists with respect to $K \times L$ and (x^0, y^0) is a saddle point.

Proof: From the assumption of the theorem follows

$$\max_K \psi(x, y^0) = \psi(x^0, y^0) = \min_L \psi(x^0, y).$$

Consequently, for any $(x, y) \in K \times L$, we have

$$\psi(x, y^0) \leqq \psi(x^0, y^0) \leqq \psi(x^0, y)$$

implying (x^0, y^0) is a saddle point. □

In general, we cannot conclude the existence of a minimax point, as we have formulated in Theorem 2.2.1, from the existence of a saddle point. This is because $\min_{L} \psi(x,y)$ must not necessarily exist for all $x \in K$ and also $\max_{K} \psi(x, y)$ must not necessarily exist for all $y \in L$. But if $M \subseteqq K$ is the domain of definition for

$$\mu(x) = \min_{L} \psi(x, y)$$

and $N \subseteqq L$ is the domain of definition for

$$\nu(y) = \max_{K} \psi(x, y)$$

then the following theorem holds.

Theorem 2.2.2. *If $\psi(x, y)$ possesses a saddle point (x^0, y^0) with respect to $K \times L$, then for $\psi(x, y)$, a minimax theorem of the form*

$$\psi(x^0, y^0) = \max_{M} \min_{L} \psi(x, y) = \min_{N} \max_{K} \psi(x, y)$$

holds.

Proof: Under the assumption, we have

$$\psi(x, y^0) \leqq \psi(x^0, y^0) \leqq \psi(x^0, y)$$

for any $(x, y) \in K \times L$. From this follows

$$\nu(y^0) = \max_{K} \psi(x, y^0) = \psi(x^0, y^0) = \min_{L} \psi(x^0, y) = \mu(x^0).$$

Thus the sets M, N are not empty because $x^0 \in M$, $y^0 \in N$.

Furthermore, we know that $\mu(x) \leqq \nu(y)$ for any $(x, y) \in M \times N$. Especially, we have

$$\mu(x) \leqq \nu(y^0) = \mu(x^0) \quad \text{for any } x \in M,$$

$$\nu(y^0) = \mu(x^0) \leqq \nu(y) \quad \text{for any } y \in N.$$

From this follows

$$\max_{M} \mu(x) = \max_{M} \min_{L} \psi(x, y) = \mu(x^0) = \nu(y^0)$$
$$= \min_{N} \nu(y) = \min_{N} \max_{K} \psi(x, y)$$

and this proves the theorem. □

From Theorem 2.2.2 it also follows immediately that the existence of a saddle point of ψ implies that there exist functions μ and ν with

$$\mu(x) = \min_{I} \psi(x, y), \qquad \nu(y) = \max_{K} \psi(x, y)$$

and the domains of definitions M and N are nonempty sets. Furthermore a theorem of duality of the form

$$\max_{M} \mu(x) = \min_{N} \nu(y)$$

is true.

In this case the problems (P) and (D) are realizable and the saddle-point coordinates x^0 and y^0 are optimal solutions of (P) and (D). On the other hand, if we have functions $\mu(x)$ and $\nu(y)$ defined on M and N and if there is a duality theorem of the form,

$$\sup_M \mu(x) = \inf_N \nu(y) = \alpha$$

where we do not demand that the problems (P) and (D) are realizable, we can immediately construct a function $\varphi(x, y)$ defined on $M \times N$ such that

$$\mu(x) = \inf_N \varphi(x, y), \qquad \nu(y) = \sup_M \varphi(x, y).$$

For this purpose, we choose a real number $\beta \in \mathbb{R}$ such that $\alpha + \beta > 0$. In case $\alpha > 0$, we put $\beta = 0$.

Then obviously, the function

$$\varphi(x, y) = \frac{\big(\mu(x) + \beta\big)\big(\nu(y) + \beta\big)}{\alpha + \beta} - \beta$$

satisfies the assumption.

In Theorem 2.2.1 we could see that from the existence of a minimax point there follows the existence of a saddle point. But we have to assume the minimax theorem in such a strong form that there exist $(x^0, y^0) \in K \times L$ giving the minmax and maxmin value $\varphi(x^0, y^0)$.

If we have a minimax theorem in the form of (A_2),

$$\sup_K \inf_L \varphi(x, y) = \inf_L \sup_K \varphi(x, y)$$

then we cannot expect that from this follows a saddle-point theorem of the form (A_3), but we will see that we get a theorem analogous to Theorem 2.2.1, if we generalize the idea of a saddle point.

Now, we assume that

$$W = \sup_K \inf_L \varphi(x, y), \qquad W^* = \inf_L \sup_K \varphi(x, y)$$

are finite values. Then the following theorem holds.

Theorem 2.2.3. *If there exists, for the function $\varphi(x, y)$, a minimax theorem in the form*

$$W = \sup_K \inf_L \varphi(x, y) = \inf_L \sup_K \varphi(x, y) = W^*$$

then for any real number $\varepsilon > 0$, there exists a pair $(x^\varepsilon, y^\varepsilon) \in K \times L$ such that

$$\varphi(x, y^\varepsilon) - \varepsilon < \varphi(x^\varepsilon, y^\varepsilon) < \varphi(x^\varepsilon, y) + \varepsilon$$

for any $(x, y) \in K \times L$.

Proof: From the assumption of the theorem follows

$$\inf_L \varphi(x, y) \leqq W \qquad \text{for any } x \in K,$$

$$\sup_K \varphi(x, y) \geqq W^* \quad \text{for any } y \in L.$$

According to the definition of infimum and supremum we find that for any $\varepsilon > 0$, there exist elements $x^\varepsilon \in K$, $y^\varepsilon \in L$ such that

$$W < \inf_L \psi(x^\varepsilon, y) + \frac{\varepsilon}{2} \leqq \psi(x^\varepsilon, y) + \frac{\varepsilon}{2} \quad \text{for any } y \in L,$$

$$W^* > \sup_K \psi(x, y^\varepsilon) - \frac{\varepsilon}{2} \geqq \psi(x, y^\varepsilon) - \frac{\varepsilon}{2} \quad \text{for any } x \in K.$$

From this follows for any $(x, y) \in K \times L$

$$\psi(x, y^\varepsilon) - \frac{\varepsilon}{2} < W^* = W < \psi(x^\varepsilon, y) + \frac{\varepsilon}{2}. \tag{2.2.2}$$

Especially from (2.2.2), we get for $(x^\varepsilon, y^\varepsilon) \in K \times L$

$$W - \frac{\varepsilon}{2} < \psi(x^\varepsilon, y^\varepsilon) < W^* + \frac{\varepsilon}{2}.$$

If we substitute (2.2.2) in the last inequality, we get

$$\psi(x, y^\varepsilon) - \varepsilon < \psi(x^\varepsilon, y^\varepsilon) < \psi(x^\varepsilon, y) + \varepsilon$$

which is the assertion of the theorem. □

If we consider

$$\mu(x) = \inf_L \psi(x, y)$$

defined on $M \subseteqq K$ and

$$\nu(y) = \sup_K \psi(x, y)$$

defined on $N \subseteqq L$, then the inequality (2.2.1) holds for this μ and ν. Now we consider the problems (P) and (D) which are not realizable. If there is a theorem of duality of the form

$$\sup_M \mu(x) = \inf_N \nu(y)$$

then this obviously is equivalent to a minimax theorem of the form

$$\sup_M \inf_L \psi(x, y) = \inf_N \sup_K \psi(x, y).$$

In connection with a generalization of a saddle point we will study more in Chapter 3.

2.3. Special Lagrangian form

In this section we treat special types of Lagrangian forms which are particularly suitable to construct well known problems of optimization, by choosing special functions as we will do in details in Chapter 3. For this purpose, the idea of conjugate functions, as they were introduced in a general way by Vogel [32], plays an important role. Using suitable special functions, in the definition of conjugate functions, we again get the idea of conjugate convex functions which we have studied in Section 1.5.

Let $\varphi(x, y)$ be a function defined on $K \times L \subseteqq \mathbb{R}^n \times \mathbb{R}^m$, $\alpha(x)$ a function defined on K and $\beta(y)$ a function defined on L. Consider the Lagrangian form

$$\psi(x, y) = \varphi(x, y) - \alpha(x) - \beta(y) \tag{2.3.1}$$

defined on $K \times L$.

Now we define the functions

$$\begin{aligned} \alpha^*(y) &= \sup_K \{\varphi(x, y) - \alpha(x)\}, \\ \beta^*(x) &= \inf_L \{\varphi(x, y) - \beta(y)\}. \end{aligned} \tag{2.3.2}$$

The suprema may exist for all $y \in N \subseteqq L$ and the infima for all $x \in M \subseteqq K$, where N and M are domains of definition for α^* and β^*. α^* and β^* are called *conjugate functions.* If $m = n$ and $\varphi(x, y) = \langle x, y\rangle$ then in case α is convex and β is concave, the functions α^* and β^* are respectively conjugate convex and conjugate concave functions. As we have done in Section 2.2, we consider

$$\mu(x) = \inf_L \psi(x, y) = \inf_L \{\varphi(x, y) - \alpha(x) - \beta(y)\} = \beta^*(x) - \alpha(x)$$

defined on $M \subseteqq K$ and

$$\nu(y) = \sup_K \psi(x, y) = \sup_K \{\varphi(x, y) - \alpha(x) - \beta(y)\} = \alpha^*(y) - \beta(y)$$

defined on $N \subseteqq L$.

Using these values in the inequality (2.2.1), we have

$$\mu(x) = \beta^*(x) - \alpha(x) \leqq \alpha^*(y) - \beta(y) = \nu(y) \tag{2.3.3}$$

for any $(x, y) \in M \times N$. This inequality constitutes a pair of dual problems:

$$\text{Find } W = \sup_M \{\beta^*(x) - \alpha(x)\}, \tag{P}$$

$$\text{Find } W^* = \inf_N \{\alpha^*(y) - \beta(y)\}. \tag{D}$$

Regarding these two problems, we have the following theorem.

Theorem 2.3.1. *The following three statements are equivalent:*

(i) *There exists a pair* $(x^0, y^0) \in K \times L$ *such that for any* $(x, y) \in K \times L$, *we have*

$$\psi(x, y^0) \leqq \psi(x^0, y^0) \leqq \psi(x^0, y).$$

(ii) *It is true that*

$$\inf_L \psi(x^0, y) = \min_L \psi(x^0, y) = \beta^*(x^0) - \alpha(x^0) = \psi(x^0, y^0),$$

$$\sup_K \psi(x, y^0) = \max_K \psi(x, y^0) = \alpha^*(y^0) - \beta(y^0) = \psi(x^0, y^0).$$

Furthermore the domains of definition M *and* N *are nonempty sets and the problems* (P) *and* (D) *are realizable and*

$$\max_M \{\beta^*(x) - \alpha(x)\} = \min_N \{\alpha^*(y) - \beta(y)\}.$$

(iii) *There exists a pair* $(x^0, y^0) \in M \times N$ *such that*

$$\varphi(x^0, y^0) = \alpha(x^0) + \alpha^*(y^0),$$
$$\varphi(x^0, y^0) = \beta(y^0) + \beta^*(x^0).$$

Points (x^0, y^0) which satisfy the statement (iii) are called *conjugate points* and they are some kind of generalization of conjugate convex points we defined in Section 1.5.

Proof: We will recognize that the saddle-point coordinates of statement (i) are optimal solutions of problems (P) and (D) and satisfy the statements (ii) as well as (iii). First we show that from statement (i) follows statement (iii).

Now, we assume for any $(x, y) \in K \times L$,

$$\varphi(x, y^0) - \alpha(x) - \beta(y^0) \leqq \varphi(x^0, y^0) - \alpha(x^0) - \beta(y^0) \leqq \varphi(x^0, y) - \alpha(x^0) - \beta(y).$$

Then follows for any $x \in K$

$$\varphi(x, y^0) - \alpha(x) \leqq \varphi(x^0, y^0) - \alpha(x^0)$$

and consequently we have

$$\alpha^*(y^0) = \sup_K \{\varphi(x, y^0) - \alpha(x)\} = \max_K \{\varphi(x, y^0) - \alpha(x)\} = \varphi(x^0, y^0) - \alpha(x^0). \tag{2.3.4}$$

Thus $y^0 \in N \subseteqq L$ and N is a nonempty set. Furthermore, from the saddle-point inequality, it follows that for any $y \in L$

$$\varphi(x^0, y^0) - \beta(y^0) \leqq \varphi(x^0, y) - \beta(y)$$

and we get

$$\begin{aligned} \varphi(x^0, y^0) - \beta(y^0) &= \min_L \{\varphi(x^0, y) - \beta(y)\} \\ &= \inf_L \{\varphi(x^0, y) - \beta(y)\} = \beta^*(x^0). \end{aligned} \tag{2.3.5}$$

Consequently $x^0 \in M \subseteqq K$ and M is a nonempty set. The relations (2.3.4) and (2.3.5) together constitute exactly the statement (iii) of the theorem

It is also easily seen that statement (ii) follows from statement (iii). Statement (iii) implies that M and N are nonempty sets. Furthermore, there follows from statement (iii) that

$$\alpha^*(y^0) = \sup_K \{\varphi(x, y^0) - \alpha(x)\} = \varphi(x^0, y^0) - \alpha(x^0) = \max_K \{\varphi(x, y^0) - \alpha(x)\}$$

and

$$\beta^*(x^0) = \inf_L \{\varphi(x^0, y) - \beta(y)\} = \varphi(x^0, y^0) - \beta(y^0) = \min_L \{\varphi(x^0, y) - \beta(y)\}.$$

Equivalently we have

$$\inf_L \psi(x^0, y) = \min_L \psi(x^0, y) = \beta^*(x^0) - \alpha(x^0) = \psi(x^0, y^0)$$

and

$$\sup_K \psi(x, y^0) = \max_K \psi(x, y^0) = \alpha^*(y^0) - \beta(y^0) = \psi(x^0, y^0).$$

From (2.3.3) follows for any $(x, y) \in M \times N$

$$\beta^*(x) - \alpha(x) \leqq \alpha^*(y^0) - \beta(y^0),$$

$$\beta^*(x^0) - \alpha(x^0) \leqq \alpha^*(y) - \beta(y).$$

These inequalities together with the statement (iii) lead to

$$\max_M \{\beta^*(x) - \alpha(x)\} = \beta^*(x^0) - \alpha(x^0) = \alpha^*(y^0) - \beta(y^0) = \min_N \{\alpha^*(y) - \beta(y)\}$$

and this proves (ii).

To complete the proof of the theorem we have to show that (ii) implies (i). Let $(x^0, y^0) \in M \times N$ be an optimal solution of the dual problems (P) and (D). Then in accordance with the statement (ii), we assume

$$\beta^*(x^0) - \alpha(x^0) = \alpha^*(y^0) - \beta(y^0). \tag{2.3.6}$$

Using the definitions of β^* and α^*, we have for any $(x, y) \in K \times L$

$$\beta^*(x^0) \leqq \varphi(x^0, y) - \beta(y)$$

and

$$\alpha^*(y^0) \geqq \varphi(x, y^0) - \alpha(x).$$

These inequalities together with (2.3.6) imply that, for any $(x, y) \in K \times L$

$$\begin{aligned} \psi(x, y^0) &= \varphi(x, y^0) - \alpha(x) - \beta(y^0) \leqq \alpha^*(y^0) - \beta(y^0) \\ &= \beta^*(x^0) - \alpha(x^0) \leqq \varphi(x^0, y) - \alpha(x^0) - \beta(y) = \psi(x^0, y^0). \end{aligned} \tag{2.3.7}$$

Furthermore, we have

$$\beta^*(x^0) - \alpha(x^0) \leqq \varphi(x^0, y^0) - \alpha(x^0) - \beta(y^0) \leqq \alpha^*(y^0) - \beta(y^0)$$

and because of (2.3.6)

$$\beta^*(x^0) - \alpha(x^0) = \psi(x^0, y^0) = \alpha^*(y^0) - \beta(y^0).$$

Substituting this equation in (2.3.7), we arrive at the saddle-point inequality and thus the theorem is proved. □

From Theorem 2.3.1 follows a simple conclusion, which we explain in some details below: Consider

$$\mathcal{L}_\alpha = \{(x, y) \in K \times N : \varphi(x, y) = \alpha(x) + \alpha^*(y)\},$$

$$\mathcal{L}_\beta = \{(x, y) \in M \times L : \varphi(x, y) = \beta(y) + \beta^*(x)\}.$$

These sets contain the community of conjugate points corresponding to the functions α and β. If the statement (iii) of Theorem 2.3.1 is true the pair $(x^0, y^0) \in M \times N$ belongs obviously to $\mathcal{L}_\alpha$ as well as $\mathcal{L}_\beta$. Consequently we have

$$\mathcal{L}_\alpha \cap \mathcal{L}_\beta \neq \emptyset.$$

On the other hand if $\mathcal{L}_\alpha \cap \mathcal{L}_\beta = \emptyset$ and $(x^0, y^0) \in \mathcal{L}_\alpha \cap \mathcal{L}_\beta$, then the pair (x^0, y^0) belongs to $M \times N$ and satisfies the statement (iii). Thus $\mathcal{L}_\alpha \cap \mathcal{L}_\beta \neq \emptyset$ if and only if the statement

(iii) of Theorem 2.3.1 is true. More generally it means that the statement $\mathscr{L}_\alpha \cap \mathscr{L}_\beta \neq \emptyset$ is equivalent to all three statements (i), (ii) and (iii) of Theorem 2.3.1.

Let us now consider the special case $m = n$. The sets K and L are in this case subsets of one and the same space $\mathbb{R}^n$ and α and β are given functions defined on K and L respectively. Furthermore, we demand that the function φ defined on $\mathbb{R}^n \times \mathbb{R}^n$ is symmetric, i.e. $\varphi(x, y) = \varphi(y, x)$. We assume that the functions α and β possess the following property: If

$$\beta^*(y) = \inf_L \{\varphi(x, y) - \beta(x)\}$$

is the conjugate function of β defined on $M \subseteqq K$ and

$$\alpha^*(y) = \sup_K \{\varphi(x, y) - \alpha(x)\}$$

is the conjugate function of α defined on $N \subseteqq L$, then we may be able to reproduce the functions α and β in the form:

$$\beta(x) = \inf_M \{\varphi(x, y) - \beta^*(y)\}$$

for all $x \in L$ and

$$\alpha(x) = \sup_N \{(x, y) - \alpha^*(y)\}$$

for all $x \in K$.

As we have done before we denote the transition to a conjugate function by a star and we put

$$[\alpha(x)]^* = \alpha^*(y), \quad [\beta(x)]^* = \beta^*(y), \quad [\varphi(x, y)]^* = \varphi(y, x).$$

Let

$$\psi(x, y) = \varphi(x, y) - \alpha(x) - \beta^*(y)$$

be a Lagrangian form defined on $K \times M$. Now we define a conjugate Lagrangian form by

$$\begin{aligned}\psi^*(x, y) &= [\varphi(x, y)]^* - [\alpha(x)]^* - [\beta^*(y)]^* = \varphi(y, x) - \alpha^*(y) - \beta^{**}(x)\\ &= \varphi(x, y) - \alpha^*(y) - \beta(x).\end{aligned}$$

Obviously the conjugate ψ^* is defined on $L \times N$ and we have $\psi^{**}(x, y) = \psi(x, y)$.

We now consider

$$\mu(x) = \inf_M \psi(x, y) = \beta(x) - \alpha(x)$$

defined on $K \subseteqq L$ and

$$\nu(y) = \sup_K \psi(x, y) = \alpha^*(y) - \beta^*(y)$$

defined on $M \subseteqq N$.

For any pair $(x, y) \in K \cap L \times M \cap N$, it is true that $\mu(x) = \beta(x) - \alpha(x) \leqq \alpha^*(y) - \beta^*(y) = \nu(y)$. In case the sets $K \cap L$ and $M \cap N$ are not empty, the formulation of the following two dual problems have a sense:

$$\text{Find } W_\psi = \sup_{K \cap L} \{\beta(x) - \alpha(x)\}, \tag{P}$$

$$\text{Find } W_\psi{}^* = \inf_{M \cap N} \{\alpha^*(y) - \beta^*(y)\}. \tag{D}$$

Regarding the conjugate Lagrangian forms we get

$$\mu^*(x) = \sup_N \psi^*(x, y) = \alpha(x) - \beta(x)$$

defined on $K \cap L$ and

$$\nu^*(y) = \inf_L \psi^*(x, y) = \beta^*(y) - \alpha^*(y)$$

defined on $M \cap N$. We also have in this case

$$\nu^*(y) = \beta^*(y) - \alpha^*(y) \leqq \alpha(x) - \beta(x) = \mu^*(x)$$

for any $(x, y) \in K \cap L \times M \cap N$ and the corresponding dual pair of problems is the following:

$$\text{Find } W^*_{\psi^*} = \sup_{M \cap N} \{\beta^*(y) - \alpha^*(y)\} \tag{P*}$$

$$\text{Find } W_{\psi^*} = \inf_{K \cap L} \{\alpha(x) - \beta(x)\}. \tag{D*}$$

Obviously, we have $W_\psi{}^* = -W^*_{\psi^*}$, $W_\psi = -W_{\psi^*}$. Furthermore it is true that

$$[\mu(x)]^* = [\beta(x)]^* - [\alpha(x)]^* = \beta^*(y) - \alpha^*(y) = \nu^*(y),$$
$$[\nu(y)]^* = [\alpha^*(y)]^* - [\beta^*(y)]^* = \alpha(x) - \beta(x) = \mu^*(x).$$

In this transformation we notice that the set K turns over to M, L to N and on the reverse M turns over to K and N to L. If we recognize this fact, then it is obviously clear, that we have (P**) = (P) and (D**) = (D).

Similar to Theorem 2.3.1, we have the following theorem.

Theorem 2.3.2. *The following three statements are equivalent:*

(i) *There exists a pair* $(x^0, y^0) \in K \cap L \times M \cap N$ *such that for any* $(x, y) \in K \times M$ *we have*

$$\psi(x, y^0) \leqq \psi(x^0, y^0) \leqq \psi(x^0, y)$$

and for any $(x, y) \in L \times N$

$$\psi^*(x^0, y) \leqq \psi^*(x^0, y^0) \leqq \psi^*(x, y^0).$$

(ii) *It is true that* $K \cap L$ *and* $M \cap N$ *are nonempty and*

$$\max_{K \cap L} \{\beta(x) - \alpha(x)\} = \min_{M \cap N} \{\alpha^*(y) - \beta^*(y)\},$$

$$\min_{K \cap L} \{\alpha(x) - \beta(x)\} = \max_{M \cap N} \{\beta^*(y) - \alpha^*(y)\}.$$

(iii) *There exists a pair* $(x^0, y^0) \in K \cap L \times M \cap N$ *such that*

$$\varphi(x^0, y^0) = \alpha(x^0) + \alpha^*(y^0),$$
$$\varphi(x^0, y^0) = \beta(x^0) + \beta^*(y^0).$$

In order to prove the Theorem 2.3.2, we do not need any essential new ideas more than what we have used in the proof of Theorem 2.3.1 and we leave it to the reader to construct the proper proof. Furthermore, analogous to the conclusion derived from Theorem 2.3.1, we can say that $\mathscr{L}_\alpha \cap \mathscr{L}_\beta \neq \emptyset$ is equivalent to the statements (i), (ii) and (iii) of Theorem 2.3.2.

So far our discussion is based on the assumption that the functions α and β have the properties, $\alpha^{**} = \alpha$ and $\beta^{**} = \beta$. Now we will find conditions under which these properties are true. Let

$$\mathscr{L}_\beta = \{(x, y) \in L \times M : \varphi(x, y) = \beta^*(y) + \beta(x)\}.$$

Now, we consider the cut for any $y \in M$ of the form

$$\mathscr{L}_{\beta^*(y)} = \{x \in L : \varphi(x, y) = \beta^*(y) + \beta(x)\}$$

and the cut for any $x \in L$ of the form

$$\mathscr{L}_{\beta(x)} = \{y \in M : \varphi(x, y) = \beta^*(y) + \beta(x)\}.$$

We assume that $\mathscr{L}_{\beta^*(y)}$ is not empty for any $y \in M$. Then there exists for any $y \in M$ at least one element $x(y) \in L$ such that

$$\varphi\big(x(y), y\big) = \beta^*(y) + \beta\big(x(y)\big).$$

Consequently, we get for any $y \in M$

$$\beta^*(y) = \inf_L \{\varphi(x, y) - \beta(x)\} = \varphi\big(x(y), y\big) - \beta\big(x(y)\big) = \min_L \{\varphi(x, y) - \beta(x)\}.$$

Thus, we can substitute infimum by minimum when we define the conjugate function β^* if $\mathscr{L}_{\beta^*(y)} \neq \emptyset$ for $y \in M$.

Theorem 2.3.3. *Let $\mathscr{L}_{\beta^*(y)}$ be a nonempty set for any $y \in M$. Then $\mathscr{L}_{\beta(x)} \neq \emptyset$ for any $x \in L$ if and only if for all $x \in L$, we have*

$$\beta(x) = \min_M \{\varphi(x, y) - \beta^*(y)\}.$$

Proof: Let us assume that for any $x \in L$, we have

$$\beta(x) = \min_M \{\varphi(x, y) - \beta^*(y)\}.$$

Then there exists for all $x \in L$ at least one element $y(x) \in M$ such that

$$\beta(x) = \min_M \{\varphi(x, y) - \beta^*(y)\} = \varphi\big(x, y(x)\big) - \beta^*\big(y(x)\big).$$

Thus $y(x) \in \mathscr{L}_{\beta(x)}$ for any $x \in L$ and $\mathscr{L}_{\beta(x)} \neq \emptyset$.

To prove the converse, we assume $\mathscr{L}_{\beta(x)} \neq \emptyset$ for any $x \in L$. Consequently for any $x \in L$, there exists $y(x) \in M$ such that

$$\varphi\big(x, y(x)\big) = \beta^*\big(y(x)\big) + \beta(x).$$

Using the definition of β^*, we have furthermore for any pair $(x, y) \in L \times M$

$$\varphi(x, y) \geqq \beta^*(y) + \beta(x).$$

Consequently

$$\varphi(x, y) - \beta^*(y) \geqq \beta(x) = \varphi(x, y(x)) - \beta^*(y(x)).$$

Thus

$$\min_M \{\varphi(x, y) - \beta^*(y)\} = \varphi(x, y(x)) - \beta^*(y(x)) = \beta(x)$$

and this proves Theorem 2.3.3. □

Similarly we can formulate such a theorem for the function α. Let us now construct a concrete example of application of Theorem 2.3.3.

Let T be a nonnegative semidefinite matrix with the property $-T = T^*$ where T^* means the transposed matrix of T. Then we have

$$0 \geqq \langle y - x, T(y - x)\rangle = \langle x, Tx\rangle + \langle y, Ty\rangle - 2\langle x, Ty\rangle.$$

Now we put

$$\varphi(x, y) = 2\langle x, Ty\rangle,$$
$$\beta(x) = \langle x, Tx\rangle$$

where β is defined on $L = \mathbb{R}^n$. For the conjugate function β^* we get

$$\beta^*(y) = \inf_{x\in\mathbb{R}^n} \{2\langle x, Ty\rangle - \langle x, Tx\rangle\}$$
$$= \inf_{x\in\mathbb{R}^n} \{-\langle y - x, T(y - x)\rangle\} + \langle y, Ty\rangle.$$

Because T is negative semidefinite the infimum will vanish for any $y \in \mathbb{R}^n$. Thus

$$\beta^*(y) = \langle y, Ty\rangle$$

defined on $M = \mathbb{R}^n$. It then follows that for any $y \in \mathbb{R}^n$

$$\mathcal{L}_{\beta^*(x)} = \{x \in \mathbb{R}^n : 2\langle x, Ty\rangle = \langle x, Tx\rangle + \langle y, Ty\rangle\} \neq \emptyset$$

for at least $x = y$ is contained in $\mathcal{L}_{\beta^*(y)}$.

Similarly, we get for any $x \in \mathbb{R}^n$

$$\mathcal{L}_{\beta(x)} = \{y \in \mathbb{R}^n : 2\langle x, Ty\rangle = \langle y, Ty\rangle + \langle x, Tx\rangle\} \neq \emptyset$$

for $y = x$ is contained in $\mathcal{L}_{\beta(x)}$. Thus the assumptions of Theorem 2.3.3 are satisfied and consequently, we have $\beta^{**} = \beta$. In fact it can easily be verified that

$$\beta^{**}(x) = \min_{y\in\mathbb{R}^n} \{2\langle x, Ty\rangle - \langle y, Ty\rangle\} = \langle x, Tx\rangle = \beta(x)$$

defined on $\mathbb{R}^n$. We can easily construct an example for which $\beta^{**} \neq \beta$. Consider

$$\varphi(x, y) = -x^2y^2,$$
$$\beta(x) = x, \quad \text{defined on } L = \{x \in \mathbb{R} : |x| \leqq 1\}.$$

Now, we have

$$\beta^*(y) = \inf_{x\in L} \{-x^2y^2 - x\} = -y^2 - 1$$

defined on $M = \mathbb{R}$. From this follows

$$\mathcal{L}_{\beta^*(y)} = \{x \in L : -x^2y^2 = -(y^2 + 1) + x\}$$

and

$$\mathscr{L}_{\beta(x)} = \{y \in \mathbb{R}\colon -x^2y^2 = -(y^2+1) + x\}.$$

It is easily that $\mathscr{L}_{\beta(x)}$ is not a none empty set for any $x \in L$; for example $\mathscr{L}_{\beta(0)} = \emptyset$. Now, by Theorem 2.3.3, $\beta^{**} \neq \beta$. In fact we get for any $x \in L$

$$\beta^{**}(x) = \min_{y\in\mathbb{R}} \{-x^2y^2 + y^2 + 1\} = 1 + (1 - x^2) \min_{y\in\mathbb{R}} y^2 = 1 .$$

Next we will demonstrate an application of Theorem 2.3.2. Consider

$$\varphi(x, y) = x^2y^2,$$

$$\alpha(x) = x^2, \quad x \in K = \{x \in \mathbb{R}\colon x > 0\},$$

$$\beta(x) = x, \quad x \in L = \{x \in \mathbb{R}\colon x \geqq 1\}.$$

The conjugate function β^* has the form,

$$\beta^*(y) = \inf_{x\geqq 1} \{x^2y^2 - x\} = \begin{cases} -\dfrac{1}{4y^2} & \text{if } 0 < y \leqq \dfrac{1}{\sqrt{2}}, \\ y^2 - 1 & \text{if } \dfrac{1}{\sqrt{2}} \leqq y \end{cases}$$

and β^* is defined on $M = K$. Furthermore we have

$$\mathscr{L}_{\beta^*(y)} = \{x \geqq 1\colon x^2y^2 = \beta^*(y) + x\}.$$

For any $y \leqq \dfrac{1}{\sqrt{2}}$, we have

$$\mathscr{L}_{\beta^*(y)} = \left\{x \geqq 1\colon x^2y^2 = -\frac{1}{4y^2} + x\right\} \neq \emptyset$$

because at least $x = \dfrac{1}{2y^2} \geqq 1$ is contained in $\mathscr{L}_{\beta^*(y)}$.

On the other hand, for any $y \geqq \dfrac{1}{\sqrt{2}}$ we have

$$\mathscr{L}_{\beta^*(y)} = \{x \geqq 1\colon x^2y^2 = y^2 - 1 + x\} \neq \emptyset$$

because the point $x = 1$ is obviously in $\mathscr{L}_{\beta^*(y)}$. Consequently $\mathscr{L}_{\beta^*(y)} \neq \emptyset$ for any $y \in M$. Now consider the set

$$\mathscr{L}_{\beta(x)} = \{y > 0\colon x^2y^2 = \beta^*(y) + x\}.$$

Then

$$\mathscr{L}_{\beta(1)} = \{y > 0\colon y^2 = \beta^*(y) + 1\} \neq \emptyset$$

because all elements y with $y \geqq \dfrac{1}{\sqrt{2}}$ belong to $\mathscr{L}_{\beta(1)}$. Furthermore, for any $x > 1$, we have

$$\mathscr{L}_{\beta(x)} = \{y > 0\colon x^2y^2 = \beta^*(y) + x\}.$$

Any y with $y^2 = \dfrac{1}{2x}$ has the property $y \leqq \dfrac{1}{\sqrt{2}}$, consequently y is contained in $\mathscr{L}_{\beta(x)}$. Thus $\mathscr{L}_{\beta(x)} \neq \emptyset$ for any $x \in L$. Now, by Theorem 2.3.3, $\beta^{**} = \beta$.

Next we find

$$\alpha^*(y) = \sup_{x>0} \{x^2y^2 - x^2\} = 0$$

and α^* is defined on $N = \{1\}$. Obviously,

$$\alpha^{**}(x) = \sup_{y=1} \{x^2y^2\} = x^2 = \alpha(x)$$

for any $x \in K$. The Lagrangian form ψ and its conjugate ψ^* will be represented by

$$\psi(x, y) = x^2y^2 - x^2 - \beta^*(y)$$

defined on $K \times K$ and

$$\psi^*(x, y) = x^2y^2 - x$$

defined on $L \times \{1\}$. We now have

$$\mu(x) = \inf_{y>0} \psi(x, y) = x - x^2, \quad x \in K \cap L = L,$$

$$\nu(y) = \sup_{x>0} \psi(x, y) = -\beta^*(y) = -y^2 + 1 = 0, \quad y \in K \cap \{1\},$$

$$\mu^*(x) = \sup_{y=1} \psi^*(x, y) = x^2 - x, \quad x \in K \cap L = L,$$

$$\nu^*(y) = \inf_{x \geqq 1} \psi^*(x, y) = \beta^*(y) = y^2 - 1, \quad y \in K \cap \{1\}.$$

The dual problems take the form:

$$\sup_{x\geqq 1} \{x - x^2\} = 0 = W_\psi, \quad x^0 = 1, \tag{P}$$

$$\inf_{y=1} \{1 - y^2\} = 0 = W_\psi{}^*, \quad y^0 = 1, \tag{D}$$

$$\sup_{y=1} \{y^2 - 1\} = 0 = W^*_{\psi^*}, \quad y^0 = 1, \tag{P*}$$

$$\inf_{x\geqq 1} \{x^2 - x\} = 0 = W_{\psi^*}, \quad x^0 = 1. \tag{D*}$$

The pair $(x^0, y^0) = (1, 1)$ satisfies the statements (i), (ii), (iii) of Theorem 2.3.2.

The statement (ii) of the Theorem 2.3.1 and 2.3.2 is a duality theorem together with the assertion about the existence of optimal solutions by which we mean realizability. But we are also able to formulate these theorems in a weaker sense by dropping the demand for realizability. For this purpose, we consider the given functions $\varphi(x, y)$, $\alpha(x)$ and $\beta(y)$ where α and β are defined on K and L respectively. Then

$$\psi(x, y) = \varphi(x, y) - \alpha(x) - \beta(y)$$

is a Lagrangian form defined on $K \times L$. Next consider

$$\alpha^*(y) = \sup_K \{\varphi(x, y) - \alpha(x)\},$$

$$\beta^*(x) = \inf_L \{\varphi(x, y) - \beta(y)\}$$

where

$$N = \left\{y \in L \colon \sup_K \{\varphi(x, y) - \alpha(x)\} < \infty\right\},$$

$$M = \left\{x \in K \colon \inf_L \{\varphi(x, y) - \beta(y)\} > -\infty\right\}$$

are the domains of definition for α^* and β^* respectively.

Now, the following theorem holds.

Theorem 2.3.4. *The following three statements are equivalent:*

(i) *For any $\varepsilon > 0$ there exists a pair $(x^\varepsilon, y^\varepsilon) \in K \times L$ such that*

$$\psi(x, y^\varepsilon) - \varepsilon < \psi(x^\varepsilon, y^\varepsilon) < \psi(x^\varepsilon, y) + \varepsilon$$

with respect to $K \times L$.

(ii) *The sets M and N are not empty and we have*

$$W = \sup_M \{\beta^*(x) - \alpha(x)\} = \inf_N \{\alpha^*(y) - \beta(y)\} = W^*.$$

(iii) *For any $\varepsilon > 0$ there exists a pair $(x^\varepsilon, y^\varepsilon) \in M \times N$ such that*

$$\varphi(x^\varepsilon, y^\varepsilon) \leqq \beta(y^\varepsilon) + \beta^*(x^\varepsilon) + \varepsilon,$$

$$\varphi(x^\varepsilon, y^\varepsilon) \geqq \alpha(x^\varepsilon) + \alpha^*(y^\varepsilon) - \varepsilon.$$

Proof: First we show that (i) implies (iii). Because of the inequality

$$\psi(x^\varepsilon, y^\varepsilon) < \psi(x^\varepsilon, y) + \varepsilon$$

we get

$$\varphi(x^\varepsilon, y) - \beta(y) > \varphi(x^\varepsilon, y^\varepsilon) - \beta(y^\varepsilon) - \varepsilon.$$

Consequently

$$\beta^*(x^\varepsilon) = \inf_L \{\varphi(x^\varepsilon, y) - \beta(y)\} \geqq \varphi(x^\varepsilon, y^\varepsilon) - \beta(y^\varepsilon) - \varepsilon.$$

Thus $M \neq \emptyset$ because x^ε is contained in M and we have the first inequality of statement (iii). In the same way, we conclude from the inequality

$$\psi(x, y^\varepsilon) < \psi(x^\varepsilon, y^\varepsilon) + \varepsilon$$

the relation

$$\varphi(x, y^\varepsilon) - \alpha(x) < \varphi(x^\varepsilon, y^\varepsilon) - \alpha(x^\varepsilon) + \varepsilon$$

for any $x \in K$, so that

$$\alpha^*(y^\varepsilon) = \sup_K \{\varphi(x, y^\varepsilon) - \alpha(x)\} \leqq \varphi(x^\varepsilon, y^\varepsilon) - \alpha(x^\varepsilon) + \varepsilon$$

and this is the second inequality of statement (iii) with $y^\varepsilon \in N \neq \emptyset$.

Next we assume statement (iii) to prove statement (ii). Statement (iii) implies that M and N are not empty. Furthermore, using the definition of conjugate functions, we have

$$\varphi(x, y) \geqq \beta^*(x) + \beta(y)$$

for any $(x, y) \in M \times L$, and

$$\varphi(x, y) \leqq \alpha(x) + \alpha^*(y)$$

for any $(x, y) \in K \times N$. From these inequalities, it follows that for any $(x, y) \in M \times N$,

$$\beta^*(x) - \alpha(x) \leqq \alpha^*(y) - \beta(y). \tag{2.3.8}$$

Furthermore, we get from the statement (iii), for any $\varepsilon > 0$ the existence of a pair $(x^\varepsilon, y^\varepsilon) \in M \times N$ such that

$$\beta^*(x^\varepsilon) - \alpha(x^\varepsilon) + \varepsilon \geqq \alpha^*(y^\varepsilon) - \beta(y^\varepsilon) - \varepsilon. \tag{2.3.9}$$

The relations (2.3.8) and (2.3.9) together imply

$$\sup_M \{\beta^*(x) - \alpha(x)\} = \inf_N \{\alpha^*(y) - \beta(y)\}$$

which is the statement (iii).

Now we have to show that from statement (ii) follows statement (i). Using the definitions of supremum and infimum, we find, for any real number $\varepsilon > 0$, elements $(x^\varepsilon, y^\varepsilon) \in M \times N$ such that

$$\sup_M \{\beta^*(x) - \alpha(x)\} < \beta^*(x^\varepsilon) - \alpha(x^\varepsilon) + \frac{\varepsilon}{2},$$

$$\inf_N \{\alpha^*(y) - \beta(y)\} > \alpha^*(y^\varepsilon) - \beta(y\ ) - \frac{\varepsilon}{2}.$$

Using the definitions of conjugate functions α^* and β^* and the assumption $W = W^*$, we get for any $(x, y) \in K \times L$

$$\begin{aligned} \varphi(x, y^\varepsilon) - \alpha(x) - \beta(y^\varepsilon) - \frac{\varepsilon}{2} &\leqq \alpha^*(y^\varepsilon) - \beta(y^\varepsilon) - \frac{\varepsilon}{2} < W^* \\ &= W < \beta^*(x^\varepsilon) - \alpha(x^\varepsilon) + \frac{\varepsilon}{2} \\ &\leqq \varphi(x^\varepsilon, y) - \beta(y) - \alpha(x^\varepsilon) + \frac{\varepsilon}{2}. \end{aligned} \tag{2.3.10}$$

Especially, we have

$$\varphi(x^\varepsilon, y^\varepsilon) - \frac{\varepsilon}{2} < W^* = W < \varphi(x^\varepsilon, y^\varepsilon) + \frac{\varepsilon}{2}$$

or

$$W^* - \frac{\varepsilon}{2} < \varphi(x^\varepsilon, y^\varepsilon) < W + \frac{\varepsilon}{2}.$$

This inequality together with (2.3.10) implies

$$\varphi(x, y^\varepsilon) - \varepsilon < \varphi(x^\varepsilon, y^\varepsilon) < \varphi(x^\varepsilon, y) + \varepsilon$$

for any $(x, y) \in K \times L$. This is in fact the statement (i) of the theorem. □

Now, we consider the special case $n = m$ and impose the restriction that the function $\varphi(x, y)$ is symmetric. Let

$$\beta^*(y) = \inf_L \{\varphi(x, y) - \beta(x)\},$$

$$\alpha^*(y) = \sup_K \{\varphi(x, y) - \alpha(x)\}$$

be the conjugate functions defined on $M \subseteqq K \subseteqq \mathbb{R}^n$ and $N \subseteqq L \subseteqq \mathbb{R}^n$ respectively.

Furthermore, we assume that for all $x \in L$, we have

$$\beta(x) = \beta^{**}(x) = \inf_M \{\varphi(x, y) - \beta^*(y)\}.$$

On refering to Lemma 1.5.2, we find that this assumption is true if β is a closed concave function defined on a closed convex set and $\varphi(x, y) = \langle x, y\rangle$. This fact will be used in Chapter 3 when we treat convex programming. If we now consider the Lagrangian form

$$\psi(x, y) = \varphi(x, y) - \alpha(x) - \beta^*(y)$$

defined on $K \times L \subseteqq \mathbb{R}^n \times \mathbb{R}^n$; then we have the following theorem:

Theorem 2.3.5. *If we assume $\beta^{**}(x) = \beta(x)$, for all $x \in L$, then the two statements given below are equivalent:*

(i) *$K \cap L \neq \emptyset$ and $M \cap N \neq \emptyset$ and we have*

$$W = \sup_{K \cap L} \{\beta(x) - \alpha(x)\} = \inf_{M \cap N} \{\alpha^*(y) - \beta^*(y)\} = W^*.$$

(ii) *For any real number $\varepsilon > 0$, there exists a pair $(x^\varepsilon, y^\varepsilon) \in K \times M$ such that*

$$\psi(x, y^\varepsilon) - \varepsilon < \psi(x^\varepsilon, y^\varepsilon) < \psi(x^\varepsilon, y) + \varepsilon$$

for any $(x, y) \in K \times L$.

Proof: We first assume that the statement (ii) is true. Then from the inequality

$$\psi(x^\varepsilon, y^\varepsilon) < \psi(x^\varepsilon, y) + \varepsilon$$

for any $y \in M$, follows

$$\varphi(x^\varepsilon, y^\varepsilon) - \beta^*(y^\varepsilon) < \varphi(x^\varepsilon, y) - \beta^*(y) + \varepsilon.$$

Because $\beta^{**} = \beta$ on L we get

$$\varphi(x^\varepsilon, y^\varepsilon) - \beta^*(y) \leqq \inf_M \{\varphi(x^\varepsilon, y) - \beta^*(y)\} + \varepsilon = \beta(x^\varepsilon) + \varepsilon. \tag{2.3.11}$$

Consequently x^ε belongs to L and thus $x^\varepsilon \in K \cap L$ i.e. $K \cap L \neq \emptyset$. From the inequality

$$\psi(x, y^\varepsilon) - \varepsilon < \psi(x^\varepsilon, y^\varepsilon)$$

for any $x \in K$ follows

$$\varphi(x, y^\varepsilon) - \alpha(x) - \varepsilon < \varphi(x^\varepsilon, y^\varepsilon) - \alpha(x^\varepsilon)$$

and consequently we get

$$\sup_K \{\varphi(x, y^\varepsilon) - \alpha(x)\} - \varepsilon = \alpha^*(y^\varepsilon) - \varepsilon \leqq \varphi(x^\varepsilon, y^\varepsilon) - \alpha(x^\varepsilon). \tag{2.3.12}$$

Thus $y^\varepsilon \in N$ and consequently $y^\varepsilon \in M \cap N$ i.e. $M \cap N \neq \emptyset$. The relations (2.3.11) and (2.3.12) together imply

$$\alpha(x^\varepsilon) + \alpha^*(y^\varepsilon) - \varepsilon \leqq \varphi(x^\varepsilon, y^\varepsilon) \leqq \beta(x^\varepsilon) + \beta^*(y^\varepsilon) + \varepsilon$$

or

$$\alpha^*(y^\varepsilon) - \beta^*(y^\varepsilon) - \varepsilon \leqq \beta(x^\varepsilon) - \alpha(x^\varepsilon) + \varepsilon. \tag{2.3.13}$$

Analogous to the inequality (2.3.8), we have on the one hand for any $(x, y) \in K \cap L \times M \cap N$

$$\beta(x) - \alpha(x) \leqq \alpha^*(y) - \beta^*(y)$$

and on the other hand, according to (2.3.13), there exists for any $\varepsilon > 0$ a pair $(x^\varepsilon, y^\varepsilon) \in K \cap L \times M \cap N$ such that

$$\alpha^*(y^\varepsilon) - \beta^*(y^\varepsilon) - \varepsilon \leqq \beta(x^\varepsilon) - \alpha(x^\varepsilon) + \varepsilon .$$

From this inequality follows immediately

$$\sup_{K \cap L} \{\beta(x) - \alpha(x)\} = \inf_{M \cap N} \{\alpha^*(y) - \beta^*(y)\}$$

which is the statement (i).

Now we assume that the statement (i) is true. Then for any $\varepsilon > 0$, there exists a pair $(x^\varepsilon, y^\varepsilon) \in K \cap L \times M \cap N$, such that

$$W - \frac{\varepsilon}{2} < \beta(x^\varepsilon) - \alpha(x^\varepsilon)$$

and

$$W^* + \frac{\varepsilon}{2} > \alpha^*(y^\varepsilon) - \beta^*(y^\varepsilon) .$$

Consequently

$$\alpha^*(y^\varepsilon) - \beta^*(y^\varepsilon) - \frac{\varepsilon}{2} < W^* = W < \beta(x^\varepsilon) - \alpha(x^\varepsilon) + \frac{\varepsilon}{2} . \tag{2.3.14}$$

Because $\beta^{**} = \beta$ for any $x \in L$, we get, using the definition of the conjugate function α^*,

$$\alpha^*(y^\varepsilon) \geqq \varphi(x, y^\varepsilon) - \alpha(x) \quad \text{for any } x \in K,$$

$$\beta(y^\varepsilon) \leqq \varphi(x^\varepsilon, y) - \beta^*(y) \quad \text{for any } y \in M .$$

If we substitute these estimates in (2.3.14), we get for any $(x, y) \in K \times M$

$$\varphi(x, y^\varepsilon) - \frac{\varepsilon}{2} < W^* = W < \varphi(x^\varepsilon, y) + \frac{\varepsilon}{2} . \tag{2.3.15}$$

Especially for $(x, y) = (x^\varepsilon, y^\varepsilon)$, we have

$$\varphi(x^\varepsilon, y^\varepsilon) - \frac{\varepsilon}{2} < W^* = W < \varphi(x^\varepsilon, y^\varepsilon) + \frac{\varepsilon}{2}$$

or

$$W^* - \frac{\varepsilon}{2} < \varphi(x^\varepsilon, y^\varepsilon) < W + \frac{\varepsilon}{2} .$$

This inequality together with the relation (2.3.15) constitute the statement (ii) of the theorem. □

2.4. Necessary and sufficient conditions for the existence of saddle points

Theorems 2.3.1 and 2.3.2 show us that saddle-point theorems, duality theorems, realizability of problems (P) and (D) and conjugate points are very closely related to one another. Besides these relationships, we are of course interested in theorems of existence of saddle points. In the context of the theory of finite (matrix) games, many scholars have studied the minimax theorems as well as saddle-point theorems. In 1922, the first minimax theorem for two person zero sum game was formulated by E. Borel in the following form:

Let T be a two by two matrix. If the sets K and L are given by

$$K = \{x \in \mathbb{R}^{2+} : \xi_1 + \xi_2 = 1\},$$

$$L = \{y \in \mathbb{R}^{2+} : \eta_1 + \eta_2 = 1\}$$

then, for $\psi(x, y) = \langle Tx, y\rangle$, it is true that

$$\max_K \min_L \psi(x, y) = \min_L \max_K \psi(x, y).$$

In 1928, J. von Neumann generalized this minimax theorem to the situation where T is any linear mapping from $\mathbb{R}^n$ into $\mathbb{R}^m$ and the sets K and L are defined by

$$K = \left\{x \in \mathbb{R}^{n+} : \sum_{i=1}^{n} \xi_i = 1\right\},$$

$$L = \left\{y \in \mathbb{R}^{m+} : \sum_{i=1}^{n} \eta_i = 1\right\}.$$

Other minimax theorems were proved by Loomis [26], Karlin [21], Fan Ky [10], Glicksberg [13] and other authors. An essential generalization of minimax theorem was found by Kakutani [19] while he was generalizing the classical theory of Brower's fixed point theorem. In 1959 Sion [31] formulated the following minimax theorem:

Let K and L be convex sets and K is compact. If $\psi(x, y)$ is quasi-concave with respect to x and upper semi-continuous at x, quasi-convex with respect to y and lower semi-continuous at y, then the minimax theorem of the form

$$\sup_K \inf_L \psi(x, y) = \inf_L \sup_K \psi(x, y)$$

holds true.

The function f is said to be *quasi-convex* if

$$x^1, x^2 \in C, \quad f(x^2) < f(x^1), \quad \lambda, \mu \in [0, 1], \quad \lambda + \mu = 1$$

implies

$$f(\lambda x^1 + \mu x^2) \leqq f(x^1),$$

and *quasi-concave* if $-f$ is quasi-convex.

Furthermore f is said to be *lower-semi-continuous at* x^0 if to each $\delta > 0$, there corresponds an open ε-neighborhood $S_\varepsilon(x^0)$ such that

$$x \in S_\varepsilon(x^0) \cap C \text{ implies } f(x) > f(x^0) - \delta,$$

and *upper semi-continuous* at x^0 if to each $\delta > 0$, there corresponds an open ε-neighborhood $S_\varepsilon(x^0)$ such that

$$x \in S_\varepsilon(x^0) \cap C \quad \text{implies} \quad f(x) < f(x^0) + \delta.$$

The proofs of all referred minimax theorems in any form use convexity and compactness. But, especially from the point of view of application of mathematical programming for practical matters, it is important to have minimax and saddle-point theorems without demanding compactness. The theory of mathematical programming for which the sets K and L are usually convex cones just means those theorems where compactness is not required. We will study this fact in the next chapter by some concrete examples of optimization.

In this section we will deal only with Kuhn-Tucker conditions [25] which ensure the existence of a saddle point and the existence of minimax theorems corresponding to Theorems 2.2.2, 2.3.1 and 2.3.2.

Theorem 2.4.1. *Let $\psi(x, y)$ be a function defined on $K \times L = \mathbb{R}^{n+} \times \mathbb{R}^{m+}$ with the property that ψ is concave in x and convex in y and Fréchet differentiable both in x and y.*

Then $(x^0, y^0) \in \mathbb{R}^{n+} \times \mathbb{R}^{m+}$ is a saddle point of ψ with respect to $K \times L$, if and only if

$$\begin{aligned} d_y\psi(x^0, y^0) &= \operatorname{grad}_y \psi(x^0, y^0) \geqq 0, \quad y^0 \in \mathbb{R}^{n+}, \quad \langle d_y\psi(x^0, y^0), y^0\rangle = 0, \\ d_x\psi(x^0, y^0) &= \operatorname{grad}_x \psi(x^0, y^0) \leqq 0, \quad x^0 \in \mathbb{R}^{n+}, \quad \langle d_x\psi(x^0, y^0), x^0\rangle = 0. \end{aligned} \tag{2.4.1}$$

Proof: First we will show that from (2.4.1) follows the saddle-point inequality. According to the assumption, $-\psi(x, y)$ is a convex function with respect to x. Now, by Theorem 1.4.10, for any $x \in \mathbb{R}^{n+}$

$$\psi(x^0, y^0) - \psi(x, y^0) \geqq -\langle d_x\psi(x^0, y^0), x - x^0\rangle.$$

Then (2.4.1) implies that

$$\psi(x, y^0) - \psi(x^0, y^0) \leqq \langle d_x\psi(x^0, y^0), x - x^0\rangle = \langle d_x\psi(x^0, y^0), x\rangle \leqq 0. \tag{2.4.2}$$

Because ψ is convex with respect to y,

$$\psi(x^0, y) - \psi(x^0, y^0) \geqq \langle d_y\psi(x^0, y^0), y - y^0\rangle = \langle d_y\psi(x^0, y^0), y\rangle \geqq 0 \tag{2.4.3}$$

for any $y \in \mathbb{R}^{m+}$. The relations (2.4.2) and (2.4.3) together constitute the saddle-point theorem.

Next, we assume that $(x^0, y^0) \in \mathbb{R}^{n+} \times \mathbb{R}^{m+}$ is a saddle point of ψ so that

$$\psi(x, y^0) \leqq \psi(x^0, y^0) \leqq \psi(x^0, y)$$

for any $(x, y) \in \mathbb{R}^{n+} \times \mathbb{R}^{m+}$. If we put

$$\chi(x) = \psi(x, y^0)$$

for $x \in \mathbb{R}^{n+}$, we have obviously

$$\chi(x) \leqq \chi(x^0) \quad \text{for} \quad x \in \mathbb{R}^{n+}$$

and the function χ possesses at least a local maximum at x^0. By Theorem 1.4.8, we have

$$(d\chi)_{x^0} = d_x\psi(x^0, y^0) \leqq 0$$

and thus

$$\langle (d\chi)_{x^0}, x^0\rangle = \langle d_x\psi(x^0, y^0), x^0\rangle = 0.$$

On the other hand, if we put

$$\chi(y) = -\psi(x^0, y)$$

then the function χ possesses at y^0 a local maximum with respect to $\mathbb{R}^{m+}$. Again by Theorem 1.4.8, we have

$$(\mathrm{d}\chi)_{y^0} = -\mathrm{d}_y\psi(x^0, y^0) \leqq 0$$

and

$$\langle(\mathrm{d}\chi)_{y^0}, y^0\rangle = -\langle\mathrm{d}_y\psi(x^0, y^0), y^0\rangle = 0.$$

This proves the theorem. □

If we specially consider on $\mathbb{R}^{n+} \times \mathbb{R}^{m+}$, the Lagrangian form

$$\psi(x, y) = \varphi(x, y) - \alpha(x) - \beta(y)$$

where α is convex, β concave and φ concave in x and convex in y, then ψ satisfies the assumptions of Theorem 2.4.1, if the functions φ, α, β are differentiable in x and y.

In this case the conditions (2.4.1) has the form

$$\begin{aligned} &\mathrm{d}_y\varphi(x^0, y^0) \geqq \mathrm{d}\beta(y^0), \quad y^0 \in \mathbb{R}^{m+}, \\ &\langle\mathrm{d}_y\varphi(x^0, y^0), y^0\rangle = \langle\mathrm{d}\beta(y^0), y^0\rangle, \\ &\mathrm{d}_x\varphi(x^0, y^0) \leqq \mathrm{d}\alpha(x^0), \quad x^0 \in \mathbb{R}^{n+}, \\ &\langle\mathrm{d}_x\varphi(x^0, y^0), x^0\rangle = \langle\mathrm{d}\alpha(x^0), x^0\rangle. \end{aligned} \tag{2.4.4}$$

It means that if the functions φ, α, β have the desired properties, mentioned above, then the conditions (2.4.4) are equivalent to the statements (i), (ii), (iii) of Theorem 2.3.1 and regarding our inquiries in Section 2.3 the conditions (2.4.4) are also equivalent to $\mathcal{L}_\alpha \cap \mathcal{L}_\beta \neq \emptyset$.

Let us continue with two remarks:

1. If we drop the demand that ψ is concave in x and convex in y, the conditions (2.4.1) and (2.4.4) remain as necessary conditions for the existence of a saddle point but they are not sufficient any more. This result follows immediately from the proof where we only used convexity to prove sufficiency.
2. We can generalize the theorem with respect to the domain of definition $\mathbb{R}^{n+} \times \mathbb{R}^{m+}$ of function ψ. Now we consider $x = (u, v) \in \mathbb{R}^{r+} \times \mathbb{R}^{s}$.

Let $\psi(u, v; y)$ be a function defined on $(\mathbb{R}^{r+} \times \mathbb{R}^{s}) \times \mathbb{R}^{m+}$ with the following properties:

(a) Either ψ is concave in (u, v), convex in y and Fréchet differentiable in all variables;
(b) or ψ possesses all partial derivatives of first order which are continuous. Then it follows that ψ is Frechet differentiable with respect to all variables and that the derivatives can be represented by gradients of ψ.

Now, we put $v = w - z$ with $w \geqq 0$, $z \geqq 0$ and consider

$$\psi(u, w - z; y) = \chi(u, w, z, y)$$

defined on $K \times L = (\mathbb{R}^{r+} \times \mathbb{R}^{s+} \times \mathbb{R}^{s+}) \times \mathbb{R}^{m+}$. Then, the conditions (2.4.1) for χ have the form

$$\operatorname{grad}_y \chi = \operatorname{grad}_y \psi \geqq 0, \quad y^0 \in \mathbb{R}^{m+},$$

$$\langle \operatorname{grad}_y \chi, y^0 \rangle = \langle \operatorname{grad}_y \psi, y^0 \rangle = 0,$$
$$\operatorname{grad}_x \chi = (\operatorname{grad}_u \chi, \operatorname{grad}_w \chi, \operatorname{grad}_z \chi) \leqq 0,$$
$$(u^0, w^0, z^0) \geqq 0,$$
$$\langle \operatorname{grad}_x \chi, u^0 \rangle + \langle \operatorname{grad}_w \chi, w^0 \rangle + \langle \operatorname{grad}_z \chi, z^0 \rangle = 0.$$

Furthermore,

$$\operatorname{grad}_u \chi = \operatorname{grad}_u \psi, \quad \operatorname{grad}_w \chi = \operatorname{grad}_v \psi, \quad \operatorname{grad}_z \chi = -\operatorname{grad}_v \psi.$$

Then, we have

$$\begin{aligned} &\operatorname{grad}_y \psi \geqq 0, \quad y^0 \in \mathbb{R}^{m+}, \quad \langle \operatorname{grad}_y \psi, y^0 \rangle = 0, \\ &(\operatorname{grad}_u \psi, \operatorname{grad}_v \psi, -\operatorname{grad}_v \psi) \leqq 0, \\ &\langle \operatorname{grad}_u \psi, u^0 \rangle + \langle \operatorname{grad}_v \psi, w^0 - z^0 \rangle = 0 \end{aligned} \tag{2.4.5}$$

or

$$\operatorname{grad}_u \psi \leqq 0, \quad \operatorname{grad}_v \psi = 0, \quad \langle \operatorname{grad}_u \psi, v^0 \rangle = 0.$$

The conditions (2.4.5) are necessary for the existence of a saddle point for χ and therefore for the existence of a saddle point for ψ. If ψ satisfies the conditions of convexity, the conditions (2.4.5) are also sufficient for a saddle point for ψ and χ.

We can also have analogous situation and similar considerations with respect to the variable y.

Now we will close this chapter with a remark concerning the generalization of the theory of duality. Inquiries of this kind, one may find for example by Golstein [14] and Göpfert [16]. All statements and proofs of Chapter 2 do not depend on the fact that we are restricting our considerations to finite dimensional spaces. Let us replace the finite dimensional space $\mathbb{R}^n$ and $\mathbb{R}^m$ by any linear Banach spaces E and F. The dual spaces of linear continuous functionals we denote E^* and F^*. Let $M \subseteqq E$ and $N \subseteqq F$ be convex cones. One can easily see that

$$M^* = \{x^* \in E^*: \langle x^*, x \rangle = x^*(x) \geqq 0; x \in M\} \subseteqq E^*,$$
$$N^* = \{y^* \in F^*: \langle y^*, y \rangle = y^*(y) \geqq 0; y \in N\} \subseteqq F^*$$

are convex cones in E^* and F^* respectively. Obviously it is

$$M = \{0\}, \quad M^* = \{x^* \in E^*: \langle x^*, 0 \rangle \geqq 0\} = E^*,$$
$$M = E, \quad M^* = \{x^* \in E^*: \langle x^*, x \rangle \geqq 0; x \in E\} = \{0\}.$$

Definition 3.1.1 defines convex cones precisely.

If we notice that for a finite dimensional space $\mathbb{R}^n$ the corresponding dual space $(\mathbb{R}^n)^*$ has the property $\mathbb{R}^n = (\mathbb{R}^n)^*$ then analogously to Section 2.1 we start from a real valued function $\psi(x, y^*)$ defined on $M \times N^* \subseteqq E \times F^*$ and repeat the considerations of the section. In Section 2.3 we start from

$$\psi(x, y^*) = \varphi(x, y^*) - \alpha(x) - \beta(y^*)$$

defined on $K \times L^* \subseteqq E \times F^*$ and receive the same equivalent statements.

3. Dual problems of optimization

3.1. Linear programming

In this chapter we will deal with special problems of optimization from the point of view of theory of duality which we have developed in Chapter 2. To begin with, we investigate the well known theory of linear programming. We will first review some essential relations which form a basis for later discussions.

Definition 3.1.1. The subset $K \subseteqq \mathbb{R}^n$ is called a *convex cone* if for any two elements $x^1, x^2 \in K$ and for any non-negative real number λ_1, λ_2, we have

$$x = \lambda_1 x^1 + \lambda_2 x^2 \in K .$$

For example, $K = \mathbb{R}^{n+} = \{x \in \mathbb{R}^n : x \geqq 0\}$ is a convex cone.

Let T be a matrix of order (m, n) and T^* the corresponding transposed matrix. Then T represents a linear mapping from $\mathbb{R}^n$ into $\mathbb{R}^m$ and T^* a linear mapping from $\mathbb{R}^m$ into $\mathbb{R}^n$.

Consider vectors $a \in \mathbb{R}^n$ and $b \in \mathbb{R}^m$. On the convex cone $K \times L = \mathbb{R}^{n+} \times \mathbb{R}^{m+}$, we consider the Lagrangian form

$$\psi_1(x, y) = -\langle y, Tx\rangle + \langle a, x\rangle + \langle b, y\rangle$$

and we put

$$\varphi_1(x, y) = -\langle y, Tx\rangle ,$$
$$\alpha_1(x) = -\langle a, x\rangle ,$$
$$\beta_1(y) = -\langle b, y\rangle .$$

Being linear, the functions α_1 and β_1 are convex as well as concave. Furthermore, $\varphi_1(x, y)$ is linear both in x and y and thus convex as well as concave in both variables. All functions $\varphi_1, \alpha_1, \beta_1$ possess partial derivatives in x and y so that Fréchet derivatives of all functions at any point in $\mathbb{R}^{n+}$ and $\mathbb{R}^{m+}$ exist.

Thus,

$$(d\alpha_1)_x = \text{grad}\,\{-\langle a, x\rangle\} = -a ,$$
$$(d\beta_1)_y = \text{grad}\,\{-\langle b, y\rangle\} = -b ,$$
$$(d\varphi_1)_x = \text{grad}_x\,\{-\langle y, Tx\rangle\} = -T^*y ,$$
$$(d\varphi_1)_y = \text{grad}_y\,\{-\langle y, Tx\rangle\} = -Tx .$$

From these relations, it follows that the function $\psi_1(x, y)$ satisfies all the assumptions of Theorem 2.4.1. Now, by theorem, if $(x^0, y^0) \in \mathbb{R}^{n+} \times \mathbb{R}^{m+}$ satisfies the conditions (2.4.4), then (x^0, y^0) is a saddle point of $\psi_1(x, y)$ with respect to $\mathbb{R}^{n+} \times \mathbb{R}^{m+}$. For this special case the conditions (2.4.4) have the form

$$\begin{aligned} Tx^0 &\leqq b, \quad x^0 \in \mathbb{R}^{n+}, \quad \langle y^0, b - Tx^0\rangle = 0, \\ Ty^0 &\geqq a, \quad y^0 \in \mathbb{R}^{m+}, \quad \langle T^*y^0 - a, x^0\rangle = 0. \end{aligned} \tag{3.1.1}$$

If (x^0, y^0) satisfies (3.1.1), then regarding Theorem 2.4.1

$$\psi_1(x, y^0) \leqq \psi_1(x^0, y^0) \leqq \psi_1(x^0, y)$$

holds for any $(x, y) \in \mathbb{R}^{n+} \times \mathbb{R}^{m+}$.

If we observe that

$$\psi_1(x, y) = \langle a, x\rangle + \langle y, b - Tx\rangle = \langle b, y\rangle + \langle a - T^*y, x\rangle$$

and substitute the expression in the saddle-point inequality, we get

$$\begin{aligned} &\langle b, y^0\rangle + \langle a - T^*y^0, x\rangle \leqq \langle b, y^0\rangle + \langle a - T^*y^0, x^0\rangle, \\ &\langle a, x^0\rangle + \langle y^0, b - Tx^0\rangle \leqq \langle a, x^0\rangle + \langle y, b - Tx^0\rangle. \end{aligned} \tag{3.1.2}$$

From (3.1.2), it follows immediately that $\sup\limits_{\mathbb{R}^{n+}} \psi_1(x, y^0)$ and $\inf\limits_{\mathbb{R}^{m+}} \psi_1(x^0, y)$ exist. Thus

$$\mu_1(x^0) = \inf_{\mathbb{R}^{m+}} \psi_1(x^0, y) = \min_{\mathbb{R}^{m+}} \psi_1(x^0, y) = \langle a, x^0\rangle + \langle y^0, b - Tx^0\rangle$$

and

$$\nu_1(y^0) = \sup_{\mathbb{R}^{n+}} \psi_1(x, y^0) = \max_{\mathbb{R}^{n+}} \psi_1(x, y^0) = \langle b, y^0\rangle + \langle a - T^*y^0, x^0\rangle.$$

Because of (3.1.1) and (3.1.2), we have

$$\mu_1(x^0) = \langle a, x^0\rangle = \psi_1(x^0, y^0) = \langle b, y^0\rangle = \nu_1(y^0) \tag{3.1.3}$$

and the sets M_1 and N_1 are not empty because they contain the saddle-point coordinates.

To define the domains of definition for μ_1 and ν_1, we consider

$$\mu_1(x) = \inf_{\mathbb{R}^{m+}} \psi_1(x, y) = \langle a, x\rangle + \inf_{\mathbb{R}^{m+}} \langle y, b - Tx\rangle$$

and

$$\nu_1(y) = \sup_{\mathbb{R}^{n+}} \psi_1(x, y) = \langle b, y\rangle + \sup_{\mathbb{R}^{n+}} \langle a - T^*y, x\rangle.$$

The infimum exists obviously if and only if $b - Tx \in \mathbb{R}^{n+}$ and it is realizable; namely the infimum is attained at $y = 0$. Thus we get

$$\mu_1(x) = \langle a, x\rangle$$

defined on

$$M_1 = \{x \in \mathbb{R}^{n+} : b - Tx \in \mathbb{R}^{m+}\}.$$

Similarly, we recognize that the supremum exists if and only if $T^*y - a \in \mathbb{R}^{n+}$ and the supremum is attained at $x = 0$. Consequently, we get

$$\nu_1(y) = \langle b, y\rangle$$

defined on

$$N_1 = \{y \in \mathbb{R}^{m+} : T^*y - a \in \mathbb{R}^{n+}\}.$$

It then follows that the conjugate functions $\alpha^*(y) = 0$ and $\beta^*(x) = 0$, regarding the inequality (2.3.3), we have for any $(x, y) \in M_1 \times N_1$

$$\mu_1(x) = \langle a, x\rangle \leqq \langle b, y\rangle = \nu_1(y). \tag{3.1.4}$$

Now, we get the dual pair of problems in the form:

$$\textit{Find } W_1 = \sup_{M_1} \langle a, x\rangle, \tag{P$_1$}$$

$$\textit{find } W_1^* = \inf_{N_1} \langle b, y\rangle. \tag{D$_1$}$$

The relations (3.1.3) and (3.1.4) show us that the problems (P_1) and (D_1) are realizable and that the saddle-point coordinates x^0 and y^0 are optimal programs. Thus

$$W_1 = \max_{M_1} \langle a, x\rangle = \langle a, x^0\rangle = \langle b, y^0\rangle = \min_{N_1} \langle b, y\rangle = W_1^*. \tag{3.1.5}$$

Furthermore, we get from (3.1.1)

$$\varphi(x^0, y^0) = -\langle y^0, Tx^0\rangle = -\langle b, y^0\rangle = \beta_1(y^0) + \beta_1^*(x^0),$$
$$\varphi(x^0, y^0) = -\langle y^0, Tx^0\rangle = -\langle a, x^0\rangle = \alpha_1(x^0) + \alpha_1^*(y^0).$$

It means that the conditions (3.1.2), (3.1.5), (3.1.1) are exactly the statements (i), (ii), (iii) of Theorem 2.3.1. Furthermore (3.1.1) means $(x^0, y^0) \in \mathscr{L}_{\alpha_1} \cap \mathscr{L}_{\beta_1}$, with

$$\mathscr{L}_{\alpha_1} = \{(x, y) \in \mathbb{R}^{n+} \times N_1 \colon \langle T^*y - a, x\rangle = 0\},$$
$$\mathscr{L}_{\beta_1} = \{(x, y) \in M_1 \times \mathbb{R}^{m+} \colon \langle y, b - Tx\rangle = 0\}.$$

Thus $\mathscr{L}_{\alpha_1} \cap \mathscr{L}_{\beta_1} \neq \emptyset$ and conversely from the relation $\mathscr{L}_{\alpha_1} \cap \mathscr{L}_{\beta_1} \neq \emptyset$ it follows that (3.1.1), (3.1.2), (3.1.5) are true.

To summarize all the results we have discussed so far, we now formulate the following theorem:

Theorem 3.1.1. *Lagrangian form*

$$\psi_1(x, y) = -\langle y, Tx\rangle + \langle a, x\rangle + \langle b, y\rangle$$

defined on convex cones $\mathbb{R}^{n+} \times \mathbb{R}^{m+}$ *produces a pair of dual problems:*

$$\textit{Find } W_1 = \max_{M_1} \langle a, x\rangle, \tag{P$_1$}$$

$$\textit{find } W_1^* = \min_{N_1} \langle b, y\rangle \tag{D$_1$}$$

with

$$M_1 = \{x \in \mathbb{R}^{n+} : b - Tx \in \mathbb{R}^{m+}\},$$
$$N_1 = \{y \in \mathbb{R}^{m+} : T^*y - a \in \mathbb{R}^{n+}\}.$$

For this problem, the following three statements are equivalent:

(i) $\psi_1(x, y^0) \leqq \psi_1(x^0, y^0) \leqq \psi_1(x^0, y)$

for any $(x, y) \in \mathbb{R}^{n+} \times \mathbb{R}^{m+}$.

(ii) (P_1) *and* (D_1) *are realizable and we have*

$$W_1 = \langle a, x^0\rangle = \max_{M_1} \langle a, x\rangle = \min_{N_1} \langle b, y\rangle = \langle b, y^0\rangle = W_1^*$$

(iii) *There exists a pair* $(x^0, y^0) \in M_1 \times N_1$ *such that*

$$\langle y^0, b - Tx^0 \rangle = 0, \quad \langle T^* y^0 - a, x^0 \rangle = 0.$$

To construct a dual pair of problems of linear programming, we started to consider a suitable Lagrangean form $\psi_1(x, y)$ defined on $K \times L = \mathbb{R}^{n+} \times \mathbb{R}^{m+}$. The domains of definition M_1 and N_1 for the constructed functions μ_1 and ν_1 which have to be optimized, were the well known linear restrictions of linear programming. If we now consider, instead of $\mathbb{R}^{n+}$ and $\mathbb{R}^{m+}$, other convex cones K and L, we arrive at further saddle-point and duality theorems as you can find them in Vogel [32]. For our purpose, we adopt the following point of view. We assume that in the restrictions of linear programming represented by the sets M_1 and N_1, the systems of inequalities $b - Tx \geqq 0$ and $T^*y - a \geqq 0$ contain some equations. Furthermore, we may weaken the assumption that the components of the vectors x and y are not negative in such a way that we admit some negative components. If we order the components of the system $b - Tx$ so that inequalities and equalities are separated out and if we further order the columns of $b - Tx$ in such a way that the first few columns contain the nonnegative components of x, then we have the following partitioned form:

$$x = (u, v) \in \mathbb{R}^r \times \mathbb{R}^s, \quad a = (a^1, a^2) \in \mathbb{R}^r \times \mathbb{R}^s, \quad r + s = n,$$

$$b = (b^1, b^2) \in \mathbb{R}^p \times \mathbb{R}^q, \quad p + q = m,$$

$$T = \begin{bmatrix} T_1 & T_2 \\ S_1 & S_2 \end{bmatrix}$$

with

$$T_1 \colon \mathbb{R}^r \to \mathbb{R}^p;\ T_1^* \colon \mathbb{R}^p \to \mathbb{R}^r,$$

$$T_2 \colon \mathbb{R}^s \to \mathbb{R}^p;\ T_2^* \colon \mathbb{R}^p \to \mathbb{R}^s,$$

$$S_1 \colon \mathbb{R}^r \to \mathbb{R}^q;\ S_1^* \colon \mathbb{R}^q \to \mathbb{R}^r,$$

$$S_2 \colon \mathbb{R}^s \to \mathbb{R}^q;\ S_2^* \colon \mathbb{R}^q \to \mathbb{R}^s.$$

Now we consider the Lagrangian form

$$\psi_2(u, v; w, z) = \varphi_2(u, v; w, z) - \alpha_2(u, v) - \beta_2(w, z)$$

with

$$\varphi_2(u, v; w, z) = -\langle w, T_1 u + T_2 v \rangle - \langle z, S_1 u + S_2 v \rangle,$$

$$\alpha_2(u, v) = -\langle a^1, u \rangle - \langle a^2, v \rangle,$$

$$\beta_2(w, z) = -\langle b^1, w \rangle - \langle b^2, z \rangle$$

defined on

$$K \times L = (\mathbb{R}^{r+} \times \mathbb{R}^s) \times (\mathbb{R}^{p+} \times \mathbb{R}^q).$$

To construct the dual pair of problems, we just find

$$\begin{aligned} \mu_2(u, v) &= \inf_L \psi_2(u, v; w, z) \\ &= \langle a^1, u \rangle + \langle a^2, v \rangle \\ &\quad + \inf_L \{\langle b^1, w \rangle + \langle b^2, z \rangle - \langle w, T_1 u + T_2 v \rangle - \langle z, S_1 u + S_2 v \rangle\} \\ &= \langle a^1, u \rangle + \langle a^2, v \rangle \\ &\quad + \inf_L \{\langle w, b^1 - T_1 u - T_2 v \rangle + \langle z, b^2 - S_1 u - S_2 v \rangle\}. \end{aligned}$$

Because $w \in \mathbb{R}^{p+}$ and $z \in \mathbb{R}^q$, the infimum exists if and only if $(u, v) \in M_2 = \{(u, v) \in \mathbb{R}^{r+} \times \mathbb{R}^s : b^1 - T_1 u - T_2 v \in \mathbb{R}^{p+};\ b^2 - S_1 u - S_2 v = 0\}$ and the infimum is attained at $w = 0$. Thus

$$\mu_2(u, v) = \langle a^1, u\rangle + \langle a^2, v\rangle$$

defined on M_2.

In the same way, we get

$$\nu_2(w, z) = \sup_K \varphi_2(u, v; w, z) = \langle b^1, w\rangle + \langle b^2, z\rangle$$
$$+ \sup_K \{\langle u, a^1 - T_1{}^*w - S_1{}^*z\rangle + \langle v, a^2 - T_2{}^*w - S_2{}^*z\rangle\}.$$

Because $u \in \mathbb{R}^{r+}$ and z is without restriction the supremum exists if and only if

$$(w, z) \in N_2 = \{(w, z) \in \mathbb{R}^{p+} \times \mathbb{R}^q : T_1{}^*w + S_1{}^*z - a^1 \in \mathbb{R}^{r+};$$
$$T_2{}^*w + S_2{}^*z - a^2 = 0\}.$$

Furthermore, the supremum is zero and we get

$$\nu_2(w, z) = \langle b^1, w\rangle + \langle b^2, z\rangle$$

defined on N_2. Regarding (1.2.4), we have

$$\mu_2(u, v) = \langle a^1, u\rangle + \langle a^2, v\rangle \leqq \langle b^1, w\rangle + \langle b^2, z\rangle = \nu_2(w, z)$$

for any $[(u, v), (w, z)] \in M_2 \times N_2$. Now the dual pair of problems take the form:

$$\text{Find } W_2 = \sup_{M_2} \{\langle a^1, u\rangle + \langle a^2, v\rangle\}, \qquad (\mathrm{P}_2)$$

$$\text{find } W_2{}^* = \inf_{N_2} \{\langle b^1, w\rangle + \langle b^2, z\rangle\}. \qquad (\mathrm{D}_2)$$

In our case

$$\mathcal{L}_{\alpha_2} = \{[(u, v), (w, z)] \in (\mathbb{R}^{r+} \times \mathbb{R}^s) \times N_2 :$$
$$\langle w, T_1 u + T_2 v\rangle + \langle z, S_1 u + S_2 v\rangle = \langle a^1, u\rangle + \langle a^2, v\rangle\},$$

$$\mathcal{L}_{\beta_2} = \{[(u, v), (w, z)] \in M_2 \times (\mathbb{R}^{p+} \times \mathbb{R}^q) :$$
$$\langle w, T_1 u + T_2 v\rangle + \langle z, S_1 u + S_2 v\rangle = \langle b^1, w\rangle + \langle b^2, z\rangle\}.$$

The condition $\mathcal{L}_{\alpha_2} \cap \mathcal{L}_{\beta_2} \neq \emptyset$ is equivalent to the statement (iii) of Theorem 2.3.1. Let $[(u^0, v^0), (w^0, z^0)]$ be an element of $\mathcal{L}_{\alpha_2} \cap \mathcal{L}_{\beta_2}$. Then this element belongs to $M_2 \times N_2$ and we have

$$\langle w^0, T_1 u^0 + T_2 v^0\rangle + \langle z^0, S_1 u^0 + S_2 v^0\rangle = \langle a^1, u^0\rangle + \langle a^2, v^0\rangle,$$
$$\langle w^0, T_1 u^0 + T_2 v^0\rangle + \langle z^0, S_1 u^0 + S_2 v^0\rangle = \langle b^1, w^0\rangle + \langle b^2, z^0\rangle.$$

But these equations are equivalent to

$$\langle u^0, T_1{}^*w^0 + S_1{}^*z^0 - a^1\rangle = 0,$$
$$\langle w^0, b^1 - T_1 u^0 - T_2 v^0\rangle = 0$$

and this is statement (iii) of Theorem 2.3.1. To summarize these results, we have the following theorem corresponding to Theorem 2.3.1.

Theorem 3.1.2. *We consider a Lagrangian form*

$$\psi_2(u, v; w, z) = \varphi_2(u, v; w, z) - \alpha_2(u, v) - \beta_2(w, z)$$

defined on $(\mathbb{R}^{r+} \times \mathbb{R}^s) \times (\mathbb{R}^{p+} \times \mathbb{R}^q)$.

This Lagrangian form produces a pair of dual problems:

$$\textit{Find } W_2 = \sup_{M_2} \{\langle a^1, u\rangle + \langle a^2, v\rangle\}; \tag{P$_2$}$$

$$\textit{find } W_2^* = \inf_{N_2} \{\langle b^1, w\rangle + \langle b^2, z\rangle\} \tag{D$_2$}$$

with

$$M_2 = \{(u, v) \in \mathbb{R}^{r+} \times \mathbb{R}^s : b^1 - T_1 u - T_2 v \in \mathbb{R}^{p+};\ b^2 - S_1 u - S_2 v = 0\},$$

$$N_2 = \{(w, z) \in \mathbb{R}^{p+} \times \mathbb{R}^q : T_1^* w + S_1^* z - a^1 \in \mathbb{R}^{r+};\ T_2^* w + S_2^* z - a^2 = 0\}.$$

Then the following three statements are equivalent:

(i) $\psi_2(u, v; w^0, z^0) \leqq \psi_2(u^0, v^0; w^0, z^0) \leqq \psi_2(u^0, v^0; w, z)$

for any $[(u, v), (w, z)] \in (\mathbb{R}^{r+} \times \mathbb{R}^s) \times (\mathbb{R}^{p+} \times \mathbb{R}^q)$.

(ii) (P_2) *and* (D_2) *are realizable and we have*

$$W_2 = \max_{M_2} \{\langle a^1, u\rangle + \langle a^2, v\rangle\} = \langle a^1, u^0\rangle + \langle a^2, v^0\rangle$$
$$= \langle b^1, w^0\rangle + \langle b^2, z^0\rangle = \min_{N_2} \{\langle b^1, w\rangle + \langle b^2, z\rangle\} = W_2^*.$$

(iii) *There exist* $[(u^0, v^0), (w^0, z^0)] \in M_2 \times N_2$, *with*

$$\langle u^0, T_1^* w^0 + S_1^* z^0 - a^1\rangle = 0,$$
$$\langle w^0, b^1 - T_1 u^0 - T_2 v^0\rangle = 0.$$

The statement (i) of Theorem 3.1.2 produces in a natural way a further saddle-point theorem. To arrive at this theorem, we consider a special case of ψ_2 by putting $S_2 = 0$. Then we get

$$\psi_3(u, v; w, z)$$
$$= -\langle w, T_1 u + T_2 v\rangle - \langle z, S_1 u\rangle + \langle a^1, u\rangle + \langle a^2, v\rangle + \langle b^1, w\rangle + \langle b^2, z\rangle$$
$$= \langle z, b^2 - S_1 u\rangle + \langle v, a^2 - T_2^* w\rangle + \chi(u, w)$$

with

$$\chi(u, w) = -\langle w, T_1 u\rangle + \langle a^1, u\rangle + \langle b^1, w\rangle.$$

The Lagrangian form ψ_3 produces a dual pair of problems:

$$\text{Find } W_3 = \sup_{M_3} \{\langle a^1, u\rangle + \langle a^2, v\rangle\}, \tag{P$_3$}$$

$$\text{find } W_3^* = \inf_{N_3} \{\langle b^1, w\rangle + \langle b^2, z\rangle\} \tag{D$_3$}$$

with

$$M_3 = \{(u, v) \in \mathbb{R}^{r+} \times \mathbb{R}^s : b^1 - T_1 u - T_2 v \in \mathbb{R}^{p+};\ b^2 - S_1 u = 0\},$$

$$N_3 = \{(w, z) \in \mathbb{R}^{p+} \times \mathbb{R}^q : T_1^* w + S_1^* z - a^1 \in \mathbb{R}^{r+};\ T_2^* w - a^2 = 0\}.$$

If $[(u^0, v^0), (w^0, z^0)] \in M_3 \times N_3$ are optimal programs of the problems (P_3) and (D_3) and if $W_3 = W_3^*$, then according to statement (i) of the theorem, these programs are saddle points of ψ_3.

Thus for any $[(u, v), (w, z)] \in (\mathbb{R}^{r+} \times \mathbb{R}^s) \times (\mathbb{R}^{p+} \times \mathbb{R}^q)$, we have

$$\begin{aligned} &\langle z^0, b^2 - S_1 u\rangle + \langle v, a^2 - T_2^* w^0\rangle + \chi(u, w^0) \\ &\leqq \langle z^0, b^2 - S_1 u^0\rangle + \langle v^0, a^2 - T_2^* w^0\rangle + \chi(u^0, w^0) \\ &\leqq \langle z, b^2 - S_1 u^0\rangle + \langle v^0, a^2 - T_2^* w\rangle + \chi(u^0, w). \end{aligned}$$

Because $a^2 - T_2^* w^0 = 0$ and $b^2 - S_1 u^0 = 0$, it follows immediately that

$$\chi(u, w^0) \leqq \chi(u^0, w^0) \leqq \chi(u^0, w)$$

for any $(u, w) \in K(\chi) \times L(\chi)$ with

$$\begin{aligned} K(\chi) &= \{u \in \mathbb{R}^{r+} : S_1 u = b^2\}, \\ L(\chi) &= \{w \in \mathbb{R}^{p+} : T_2^* w = a^2\}. \end{aligned}$$

Consequently if the problems (P_3) and (D_3) possess optimal programs $[(u^0, v^0), (w^0, z^0)] \in M_3 \times N_3$ and if $W_3 = W_3^*$, then for χ there exists a saddle point with respect to $K(\chi)$, $L(\chi)$ where $(u^0, w^0) \in K(\chi) \times L(\chi)$ are saddle-point coordinates. On the other hand it is possible to prove that the problems (P_3) and (D_3) are solvable so far as there exists a saddle point $(u^0, w^0) \in K(\chi) \times L(\chi)$ for χ. To establish this result, it is sufficient to show that the sets M_3, N_3 are not empty. If $M_3 \neq \emptyset$ and $N_3 \neq \emptyset$, it follows that, for any $[(u, v), (w, z)] \in M_3 \times N_3$,

$$\langle a^1, u\rangle + \langle a^2, v\rangle \leqq \langle b^1, w\rangle + \langle b^2, z\rangle.$$

But this inequality means that the supremum and infimum for the problems (P_3) and (D_3) respectively exist.

To show that M_3 and N_3 are not empty under the condition that a saddle point exists for χ, we prove the following lemma.

Lemma 3.1.1. (Farkas lemma). *Let T be a matrix of order (m, n) and b a given vector in $\mathbb{R}^m$. The set*

$$\{x \in \mathbb{R}^{n+} : Tx = b\}$$

is not empty if and only if

$$\langle b, y\rangle \geqq 0 \quad \textit{for any } y \in \mathbb{R}^m$$

with $T^* y \geqq 0$.

Proof: Suppose $\{x \in \mathbb{R}^{n+} : Tx = b\} \neq \emptyset$ and let x be any element of this set. Then we have, for any $y \in \mathbb{R}^m$ with $T^* y \geqq 0$

$$0 \leqq \langle x, T^* y\rangle = \langle Tx, y\rangle = \langle b, y\rangle.$$

Now we assume $\langle b, y\rangle \geqq 0$ for any $y \in \mathbb{R}^m$ with $T^* y \geqq 0$. Under the assumption

$$\{x \in \mathbb{R}^{n+} : Tx = b\} = \emptyset,$$

we can construct a contradiction.

Consider the set

$$T(\mathbb{R}^{n+}) = \{y \in \mathbb{R}^m : y = Tx;\, x \in \mathbb{R}^{n+}\}.$$

The set $T(\mathbb{R}^{n+})$ is convex and closed. Under the assumption that $\{x \in \mathbb{R}^{n+} : Tx = b\}$ is empty, the element b is not contained in $T(\mathbb{R}^{n+})$. Then according to the separation Theorem 1.2.2, there exists a hyperplane

$$H = \{y \in \mathbb{R}^m : \langle y^0, y \rangle - \gamma = 0\}$$

which separates the set $T(\mathbb{R}^{n+})$ from $\{b\}$ strictly. Thus,

$$\langle y^0, y \rangle - \gamma > 0 \quad \text{for any } y \in T(\mathbb{R}^{n+}),$$
$$\langle y^0, b \rangle - \gamma < 0.$$

From this follows for any $y \in T(\mathbb{R}^{n+})$

$$\langle y^0, Tx \rangle > \gamma.$$

Especially for $x = 0$, we have $\gamma < 0$. Let $\sigma > 0$ be any real number and e^i the vector in $\mathbb{R}^n$ with all components zero except the ith which is one. Then we get, for any $\sigma > 0$ and $x = \sigma e^i \in \mathbb{R}^{n+}$,

$$\langle y^0, T(\sigma e^i) \rangle = \sigma \langle T^* y^0, e^i \rangle = \sigma [T^* y^0]_i > \gamma.$$

Because γ is negative, this is only possible if $[T^* y^0]_i \geqq 0$, $i = 1, 2, \ldots, n$ i.e. $T^* y^0 \geqq 0$. This means that y^0 is an element in $\mathbb{R}^m$ such that $T^* y^0 \geqq 0$ and on the other hand according to the separation theorem, we have

$$\langle y^0, b \rangle < \gamma < 0$$

which contradicts the assumption. □

From Lemma 3.1.1, it follows that the system

$$\begin{aligned} &T_1 u + T_2 v \leqq b^1, \\ &S_1 u + S_2 v = b^2, \\ &(u, v) \in \mathbb{R}^{r+} \times \mathbb{R}^s \end{aligned} \tag{3.1.6}$$

is solvable if and only if

$$\langle b^1, w \rangle + \langle b^2, z \rangle \geqq 0 \quad \text{for any } (w, z) \in \mathbb{R}^{p+} \times \mathbb{R}^q$$

with

$$\begin{aligned} &T_1^* w + S_1^* z \geqq 0, \\ &T_2^* w + S_2^* z = 0. \end{aligned} \tag{3.1.7}$$

In fact, if we consider the system,

$$\begin{bmatrix} T_1 & T_2 & -T_2 & I \\ S_1 & S_2 & -S_2 & 0 \end{bmatrix} \begin{bmatrix} u \\ v^1 \\ v^2 \\ t \end{bmatrix} = \begin{bmatrix} b^1 \\ b^2 \end{bmatrix}, \quad \begin{bmatrix} u \\ v^1 \\ v^2 \\ t \end{bmatrix} \geqq 0, \tag{3.1.8}$$

its solvability is equivalent to the solvability of (3.1.6), i.e. every solution of (3.1.8) produces a solution for (3.1.6) with $v = v^1 - v^2$. On the other hand, if we consider a solution of (3.1.6), then we get a solution for (3.1.8) by putting $t = b^1 - T_1 u - T_2 v$ where v is represented in the form $v = v^1 - v^2$ with $v^1, v^2 \in \mathbb{R}^{s+}$.

The solvability conditions for (3.1.8), according to the Lemma 3.1.1 are now given: For any $(w, z) \in \mathbb{R}^{p+} \times \mathbb{R}^q$, satisfying

$$\begin{bmatrix} T_1^* & S_1^* \\ T_2^* & S_2^* \\ -T_2^* & -S_2^* \\ I & 0 \end{bmatrix} [w, z] = \begin{bmatrix} T_1^*w + S_1^*z \\ T_2^*w + S_2^*z \\ -T_2^*w - S_2^*z \\ w \end{bmatrix} \geqq 0$$

we have

$$\langle b^1, w\rangle + \langle b^2, z\rangle \geqq 0$$

which is exactly the condition (3.1.7).

From (3.1.6) and (3.1.7), by considering special cases, we immediately deduce the following statements:

Case I $T_2 = 0,\ S_1 = 0,\ S_2 = 0$.

The system $T_1u \leqq b$, $u \in \mathbb{R}^{r+}$ is solvable if and only if $\langle b^1, w\rangle \geqq 0$ for any $w \in \mathbb{R}^{p+}$ with $T_1^*w \geqq 0$. (3.1.9)

Case II $T_1 = 0,\ S_1 = 0,\ S_2 = 0$.

The system $T_2v \leqq b^1$, $v \in \mathbb{R}^s$ is solvable if and only if $\langle b^1, w\rangle \geqq 0$ for any $w \in \mathbb{R}^{p+}$ with $T_2^*w = 0$. (3.1.10)

Case III $T_1 = 0,\ T_2 = 0,\ S_1 = 0$.

The system $S_2v = b^2$, $v \in \mathbb{R}^s$ is solvable if and only if $\langle b^2, z\rangle = 0$ for any $z \in \mathbb{R}^q$ with $S_2^*z = 0$. (3.1.11)

Case IV $T_1 = 0,\ T_2 = 0,\ S_2 = 0$.

The system $S_1u = b^2$, $u \in \mathbb{R}^{r+}$ is solvable if and only if $\langle b^2, z\rangle \geqq 0$ for any $z \in \mathbb{R}^q$ with $S_1^*z \geqq 0$. (3.1.12)

Now, we return to the assertion that from the existence of a saddle point for χ it follows that the sets M_3 and N_3 are not empty. We restrict our considerations to prove this statement for M_3; as for N_3 the situation is analogous and the proof is similar. If $u^0 \in K(\chi)$ is a saddle-point coordinate for χ, then $M_3 \neq \emptyset$ if there exists $v \in \mathbb{R}^s$ such that

$$T_2v \leqq b^1 - T_1u^0. \tag{3.1.13}$$

According to (3.1.10), the system (3.1.13) is solvable if $\langle b^1 - T_1u, w\rangle \geqq 0$ for any $w \in \mathbb{R}^{p+}$ with $T_2^*w = 0$. Now we take these w and put $\tilde{w} = w + w^0$ where w^0 is the second saddle-point coordinate of χ. Together with w and w^0, $\tilde{w}$ also belongs to $L(\chi)$. Then, we have, regarding the saddle-point inequality,

$$\begin{aligned} \chi(x^0, w^0) &= -\langle w^0, T_1u^0\rangle + \langle a^1, u^0\rangle + \langle b^1, w^0\rangle \\ &\leqq -\langle w + w^0, T_1u^0\rangle + \langle a^1, u^0\rangle + \langle b^1, w + w^0\rangle = \chi(u^0, \tilde{w}). \end{aligned}$$

From this follows

$$0 \leqq \langle w, b^1 - T_1u^0\rangle.$$

Consequently, (3.1.13) is solvable.

If we remember the well known duality theorem of linear programming which says that $M_3 \neq \emptyset$ and $N_3 \neq \emptyset$ are equivalent to the existence of optimal programs for (P_3)

and (D_3), we can even say that from the existence of a saddle point for χ follows the existence of optimal programs for the dual pair (P_3) and (D_3).

To summarize these results we formulate the following theorem.

Theorem 3.1.3. *If there exist optimal programs* $[(u^0, v^0), (w^0, z^0)] \in M_3 \times N_3$ *for the dual pair* (P_3) *and* (D_3) *and if we have* $W_3 = W_3^*$, *then*

$$\chi(u, w^0) \leqq \chi(u^0, w^0) \leqq \chi(u^0, w)$$

for any $(u, w) \in K(\chi) \times L(\chi)$.

Conversely if a saddle point for χ *exists with respect to* $K(\chi) \times L(\chi)$, *then the sets* M_3 *and* N_3 *are not empty and the problems* (P_3) *and* (D_3) *possess optimal programs.*

A suitable special case of the problems (P_3) and (D_3) will lead us to the minimax theorem in the theory of two person zero sum game. Now, we will present this special case to be followed by the interpretation from the point of view of theory of games. Let e_n and e_m respectively denote vectors of $\mathbb{R}^n$ and $\mathbb{R}^m$ with the property that all their components are one.

Now we put

$$T_1 = -T : \mathbb{R}^n \to \mathbb{R}^m, \quad T_1^* = -T^* : \mathbb{R}^m \to \mathbb{R}^n,$$
$$T_2 = e_m^* : \mathbb{R}^1 \to \mathbb{R}^m, \quad T_2^* = e_m : \mathbb{R}^m \to \mathbb{R}^1,$$
$$S_1 = e_n : \mathbb{R}^n \to \mathbb{R}^1, \quad S_1^* = e_n^* : \mathbb{R}^1 \to \mathbb{R}^n.$$

Furthermore, we consider

$$(x, \xi) \in \mathbb{R}^n \times \mathbb{R}^1, \quad (y, \eta) \in \mathbb{R}^m \times \mathbb{R}^1,$$
$$(b^1, b^2) = (0, 1) \in \mathbb{R}^m \times \mathbb{R}^1, \quad (a^1, a^2) = (0, 1) \in \mathbb{R}^n \times \mathbb{R}^1.$$

Then we get

$$\varphi_4(x, \xi; y, \eta) = \langle y, Tx - \xi e_m \rangle - \eta \langle x, e_n \rangle,$$
$$\alpha_4(x, \xi) = -\xi,$$
$$\beta_4(y, \eta) = -\eta.$$

The Lagrangian form ψ_4 is defined on $K \times L = (\mathbb{R}^{n+} \times \mathbb{R}^1) \times (\mathbb{R}^{m+} \times \mathbb{R}^1)$ in the form

$$\begin{aligned}\psi_4(x, \xi; y, \eta) &= \langle y, Tx - \xi e_m^* \rangle - \eta \langle x, e_n \rangle + \xi + \eta \\ &= \langle Tx, y \rangle + \eta(1 - \langle x, e_n \rangle) + \xi(1 - \langle y, e_m \rangle) \\ &= \eta(1 - \langle x, e_n \rangle) + \xi(1 - \langle y, e_m \rangle) + \chi(x, y)\end{aligned}$$

with

$$\chi(x, y) = \langle Tx, y \rangle.$$

The dual pair of problems now appears in the following form:

$$\text{Find } W_4 = \sup_{M_4} \xi, \tag{P$_4$}$$

where $M_4 = \{(x, \xi) \in \mathbb{R}^{n+} \times \mathbb{R}^1 : Tx - \xi e_m^* \in \mathbb{R}^{m+};\ 1 - \langle x, e_n \rangle = 0\}$.

$$\text{Find } W_4^* = \inf_{N_4} \eta, \tag{D$_4$}$$

where $N_4 = \{(y, \eta) \in \mathbb{R}^{m+} \times \mathbb{R}^1 : \eta e_n^* - T^* y \in \mathbb{R}^{n+};\ \langle y, e_m \rangle - 1 = 0\}$.

Now consider the function $\chi(x, y) = \langle Tx, y\rangle$ on $K(\chi) \times L(\chi)$ where

$$K(\chi) = \{x \in \mathbb{R}^{n+} : \langle x, e_n\rangle = 1\},$$
$$L(\chi) = \{y \in \mathbb{R}^{m+} : \langle y, e_m\rangle = 1\}.$$

Because ξ and η are real numbers without any restrictions, the sets M_4 and N_4 are always not empty and thus the problems (P_4) and (D_4) are always solvable.

To interpret this problem from the point of view of the theory of games of strategy, we consider a two-person zero sum game involving two players A and B with conflicting interests. The outcome of such games is usually determined partly by the intelligence and skills of one side and partly by those of the opposing side. The player A has the possibility to choose one of the n decisions while B can make one of the m decisions in the game against A. Every possible decision made by A or B, we call a *pure strategy*. It means that the players A and B are able to choose one of n and m pure strategies respectively. If the choice of A is the kth pure strategy while the choice of B is the ith pure strategy, we denote the effect of the decision against each other by t_{ik} where we understand the value t_{ik} as profit for A or as damage for B. If one player chooses the kth of his pure strategies, we can interpret this act in such a way that he has chosen the kth pure strategy with probability one and each of the remaining strategies with probability zero. It is natural to understand that a player is in a better position if he choose instead of a pure strategy a so called *mixed strategy* which means a probability distribution over the set of all pure strategies. From this point of view, $K(\chi)$ and $L(\chi)$ are the sets of mixed strategies of A and B respectively. A component ξ_k of vector $x \in K(\chi)$ is the probability that the player A chooses the kth pure strategy. Of course, the pure strategies are contained in $K(\chi)$ and $L(\chi)$ in the form of unit vectors. The matrix $T = [t_{ik}]$ of order (m, n), known as the pay-off matrix, describes the results of the game in the form of profit for one player and damage for the other corresponding to different pairs of pure strategies. We denote the pure strategies by $e_n^k \in K(\chi)$ and $e_m^i \in L(\chi)$, $k = 1, \ldots, n$; $i = 1, \ldots, m$, which are the unit vectors in $\mathbb{R}^n$ and $\mathbb{R}^m$ respectively. We have, for the pay-off for A,

$$\chi(e_n^k, e_m^i) = \langle Te_n^k, e_m^i\rangle = t_{ik},$$

if both players used pure strategies e_n^k and e_m^i. In general,

$$\chi(x, y) = \langle Tx, y\rangle$$

represents the expected value of pay-off for A corresponding to the choice of mixed strategies $(x, y) \in K(\chi) \times L(\chi)$. The interest of both players is to choose such optimal strategies $(x^0, y^0) \in K(\chi) \times L(\chi)$ that the expected value of pay-off is a maximum for player A and a minimum for player B. For this special case of theory of games, Theorem 3.1.2 takes the following form:

Theorem 3.1.4. *The Lagrangian form*

$$\psi_4(x, \xi; y, \eta) = \langle Tx, y\rangle + \eta(1 - \langle x, e_n\rangle) + \xi(1 - \langle y, e_m\rangle)$$

defined on $(\mathbb{R}^{n+} \times \mathbb{R}^1) \times (\mathbb{R}^{m+} \times \mathbb{R}^1)$, *produces a pair of dual problems*

$$\text{find } W_4 = \sup_{M_4} \xi; \qquad (P_4)$$

$$\text{find } W_4^* = \inf_{N_4} \eta \qquad (D_4)$$

where

$$M_4 = \{(x, \xi) \in \mathbb{R}^{n+} \times \mathbb{R}^1 : Tx - \xi e_m^* \in \mathbb{R}^{m+};\ 1 - \langle x, e_n \rangle = 0\},$$
$$N_4 = \{(y, \eta) \in \mathbb{R}^{m+} \times \mathbb{R}^1 : \eta e_n^* - T^* y \in \mathbb{R}^{n+};\ \langle y, e_m \rangle - 1 = 0\}.$$

Then, the following statements are equivalent.

(i) *There exists a point*

$$[(x^0, \xi_0), (y^0, \eta_0)] \in (\mathbb{R}^{n+} \times \mathbb{R}^1) \times (\mathbb{R}^{m+} \times \mathbb{R}^1)$$

such that

$$\psi_4(x, \xi; y^0, \eta^0) \leqq \psi_4(x^0, \xi_0; y^0, \eta_0) \leqq \psi_4(x^0, \xi_0, y, \eta)$$

for any

$$[(x, \xi), (y, \eta)] \in (\mathbb{R}^{n+} \times \mathbb{R}^1) \times (\mathbb{R}^{m+} \times \mathbb{R}^1).$$

(ii) *The problems* (P_4) *and* (D_4) *are realizable and*

$$W_4 = \max_{M_4} \xi = \xi_0 = \eta_0 = \min_{N_4} \eta = W_4^*.$$

(iii) *There exists a point* $[(x^0, \xi_0), (y^0, \eta_0)] \in M_4 \times N_4$ *such that*

$$\langle x^0, \eta_0 e_n^* - T^* y^0 \rangle = 0,$$
$$\langle y^0, T x^0 - \xi_0 e_m^* \rangle = 0.$$

If one of the statements (i), (ii), (iii) is true, we have according to Theorem 3.1.3,

$$\langle Tx, y^0 \rangle \leqq \langle Tx^0, y^0 \rangle \leqq \langle Tx^0, y \rangle \tag{3.1.14}$$

with respect to all mixed strategies $(x, y) \in K(\chi) \times L(\chi)$. Furthermore, from (ii) and (iii) follows

$$\langle Tx^0, y^0 \rangle = \xi_0 = W_4 = W_4^* = \eta_0.$$

The common value $W_4 = W_4^* = \lambda$ is called the *value of the game*. Thus the value of the game is equal to the level of the saddle point of $\langle Tx, y \rangle$. The interpretation of these facts now follows.

If A chooses the mixed strategy x^0, then according to (3.1.14) we have $\langle Tx^0, y \rangle \geqq \lambda$ for any mixed strategy y for B. Thus the pay-off for A is at least λ; and the player A ensured that the expected value of his pay-off is not less than the value of the game. If B chooses the mixed strategy y^0, then we have $\langle Tx, y^0 \rangle \leqq \lambda$ for any mixed strategy x for A. Thus the damage for B is at most λ and the player B ensured that the expected value of his damage is not more than the value λ of the game. From this point of view, x^0 and y^0 are optimal mixed strategies for the players A and B.

The matrix T is a linear continuous bounded operator. Furthermore, for $(x, y) \in K(\chi) \times L(\chi)$, we have $\|x\| \leqq 1$ and $\|y\| \leqq 1$. Consequently,

$$|\langle Tx, y \rangle| \leqq \|T\| \, \|x\| \, \|y\| \leqq \|T\|$$

where we define

$$\|T\| = \sup_{\|x\| \leqq 1} \|Tx\|.$$

Then it follows that

$$-\|T\| \leqq \langle Tx, y \rangle \leqq \|T\|.$$

Thus there exist

$$\sup_{K(\chi)} \langle Tx, y\rangle, \inf_{L(\chi)} \langle Tx, y\rangle$$

for any $y \in L(\chi)$ and $x \in K(\chi)$ respectively.

Regarding Theorems 2.2.1 and 2.2.2, the saddle-point inequality (3.1.14) is equivalent to

$$\max_{K(\chi)} \min_{L(\chi)} \langle Tx, y\rangle = \min_{L(\chi)} \max_{K(\chi)} \langle Tx, y\rangle .$$

But this is exactly the minimax theorem of J. von Neumann.

3.2. Linear programming in dual paired spaces

The speciality of linear programming in Section 3.1 is the choise of linear spaces as finite dimensional spaces $\mathbb{R}^n$ and $\mathbb{R}^m$. In this case a linear operator T mapping from $\mathbb{R}^n$ into $\mathbb{R}^m$ can be represented in form of a matrix of order (m, n). The corresponding transposed matrix T^* mapping from $\mathbb{R}^m$ into $\mathbb{R}^n$ is the dual operator to T and can be characterized by

$$\langle Tx, y\rangle = \langle x, T^*y\rangle \tag{3.2.1}$$

for any $(x, y) \in D(T) \times D(T^*)$, $D(T)$ and $D(T^*)$ being domain of definition of T and T^* respectively. It seems quite natural for generalization of linear programming to replace the finite dimensional spaces $\mathbb{R}^n$ and $\mathbb{R}^m$ by any linear Banach spaces E and F. Investigations of this kind one may find by Kretschmer [24].

For constructing the theory of linear programming in any Banach spaces we first make some introducing remarks. Let E and F be any linear Banach spaces and E^*, F^* the corresponding dual spaces of linear continuous functionals defined on E and F respectively. Because the spaces E^* and F^* are linear Banach spaces also it makes sense to consider the totality of linear continuous functionals defined on E^* and F^* respectively. We denote $(E^*)^* = E^{**}$ and $(F^*)^* = F^{**}$. Consider any fixed element $x \in E$. Then $\varphi_x(x^*) = x^*(x)$ is a linear continuous functional defined on E^* thus $\varphi_x \in E^{**}$. It now follows $E \subseteqq E^{**}$. A linear Banach space with the property $E^{**} = E$ we call reflexive.

A sequence $\{x_n\} \subseteqq E$ is said to be *weak converging* to $x_0 \in E$ if for all $x^* \in E^*$ it holds

$$\lim_{n\to\infty} \langle x^*, x_n\rangle = \langle x^*, x_0\rangle$$

and a sequence $\{x_n^*\} \subseteqq E^*$ is *weak converging* to $x_0^* \in E^*$ if for all $x \in E$ it follows

$$\lim_{n\to\infty} \langle x_n^*, x\rangle = \langle x_0^*, x\rangle .$$

Besides weak convergence we consider the usual convergence of a sequence $\{x_n\}$ with respect to the norm $\|\cdot\|$ in E. For further considerations we assume E and F being reflexive. Now let $P \subseteqq E$ be a closed convex cone. Obviously the set

$$P^* = \{x^* \in E^*: \langle x^*, x\rangle \in \mathbb{R}^+ \text{ for all } x \in P\} \subseteqq E^*$$

is a convex cone in E^*. We call P^* the dual convex cone. Now it is possible to construct

$$(P^*)^* = P^{**} = \{x \in E: \langle x^*, x\rangle \in \mathbb{R}^+ \text{ for all } x^* \in P^*\} \subseteqq E .$$

Lemma 3.2.1. *If P is a closed convex cone then $P^{**} = P$.*

Proof: Let $x_0 \in E$ and $\langle x^*, x_0\rangle \in \mathbb{R}^+$ for all $x^* \in P^*$. Now we prove $x_0 \in P$. If we assume $x_0 \notin P$ because P being closed it follows by Theorem 1.2.2 the existence of a hyperplane

$$H = \{x \in E : \langle x_0^*, x\rangle - \gamma = 0\}$$

such that

$$\langle x_0^*, x_0\rangle < \inf_{x \in P} \langle x_0^*, x\rangle . \tag{3.2.2}$$

$0 \in P$ implies that $\inf_{x\in P} \langle x_0^*, x\rangle \leqq 0$. In case $\inf_{x\in P} \langle x_0^*, x\rangle < 0$ it follows the existence of an element $\bar{x} \in P$ such that $\langle x_0^*, \bar{x}\rangle < 0$. Because for any $\lambda \in \mathbb{R}^+$ holds $\lambda\bar{x} \in P$ it follows fo great λ, $\langle x_0^*, \lambda x\rangle < \langle x_0^*, x_0\rangle$. With respect to (3.1.2) this is not possible, thus

$$\inf_{x\in P} \langle x_0^*, x\rangle = 0 \leqq \langle x_0^*, x\rangle \text{ for all } x \in P .$$

Then we can see that $x_0^* \in P^*$ and (3.1.2) implies $\langle x_0^*, x_0\rangle < 0$. But this is a contradiction to the assumption $\langle a^*, x_0\rangle \in \mathbb{R}^+$ for all $x^* \in P^*$. Thus $x_0 \in P$.

Now it follows

$$P^{**} = \{x \in E : \langle x^*, x\rangle \in \mathbb{R}^+ \text{ for all } x^* \in P^*\} \subseteqq P .$$

On the other hand for an element $x \in P$ we get for all $x^* \in P^*$, $\langle x^*, x\rangle \geqq 0$ thus $P \subseteqq P^{**}$. Consequently it is $P = P^{**}$. Now let E and F be reflexive Banach spaces and E^*, F^* the corresponding dual spaces. We consider any closed convex cones $P \subseteqq E$ and $Q \subseteqq F$ and denote the dual convex cones by P^* and Q^* respectively. For elements of the different spaces we use the following notation: $x \in E$, $x^* \in E^*$, $y \in F$, $y^* \in F^*$. Now let T be a linear bounded operator mapping from E into F. Then the corresponding dual linear operator T^* mapping from F^* into E^* can be defined by

$$\langle y^*, Tx\rangle = \langle T^*y^*, x\rangle$$

for all $(x, y^*) \in D(T) \times D(T^*)$. Here $D(T)$ and $D(T^*)$ denote the domain of definition of T and T^* respectively. Consider elements $a^* \in E^*$ and $b \in F$. One the convex cone $P \times Q^*$, we consider the Lagrangian form

$$\psi(x, y^*) = -\langle y^*, Tx\rangle + \langle a^*, x\rangle + \langle y^*, b\rangle .$$

Analogously to Section 2.3 we construct dual optimization problems:

$$\inf_{Q^*} \psi(x, y^*) = \langle a^*, x\rangle + \inf_{Q^*} \langle y^*, b - Tx\rangle = \begin{cases} \langle a^*, x\rangle, & x \in Z, \\ -\infty, & x \notin Z. \end{cases}$$

Z denotes the set of admissible points and is defined by

$$Z = \{x \in P : b - Tx \in Q\} .$$

Furthermore we get

$$\sup_{P} \psi(x, y^*) = \langle y^*, b\rangle + \sup_{P} \langle a^* - T^*y^*, x\rangle = \begin{cases} \langle y^*, b\rangle, & y^* \in Z^*, \\ \infty, & y^* \notin Z^*. \end{cases}$$

The set Z^* of admissible points we get in form of

$$Z^* = \{y^* \in Q^* : T^*y^* - a^* \in P^*\} .$$

Now we get the dual pair of problems in the form:

$$W = \sup_{Z} \langle a^*, x \rangle, \tag{P}$$

$$W^* = \inf_{Z^*} \langle y^*, b \rangle. \tag{D}$$

The problem (P) formed by elements $[T, P, Q, a^*, b]$ we call a *primary program*. Let us consider the following transformation:

$$\begin{array}{lll} T \to -T^*, & P \to Q^*, & Q \to P^*, \\ a^* \to -b, & b \to -a^*. & \end{array} \tag{3.2.3}$$

Then we use the notation

$$[T, P, Q, a^*, b]^{\circledast} = [-T^*, Q^*, P^*, -b, -a^*]$$

and call this quintupled the *dual program*. Applying the transformation (3.2.3) twice with respect to Lemma 3.2.1 we reproduce the primary program i.e.

$$([T, P, Q, a^*, b]^{\circledast})^{\circledast} = [-T^*, Q^*, P^*, -b, -a^*]^{\circledast} = [T, P, Q, a^*, b].$$

Thus we realize the principle of duality.

Definition 3.2.1. We call the primary program $[T, P, Q, a^*, b]$ *consistent* if $Z \neq \emptyset$; the primary program is said to be *subconsistent* if $b \in \overline{T(P) + Q}$; analogous we call the dual program *consistent* if $Z^* \neq \emptyset$ and the dual program is *subconsistent* if $a^* \in \overline{T^*(Q^*) - P^*}$.

Here $\overline{T(P) + Q}$ and $\overline{T^*(Q^*) - P^*}$ denote the closure of $T(P) + Q$ and $T^*(Q^*) - P^*$ respectively.

We call a sequence $\{x_n\} \subseteqq P$ *admissible* if there exists a sequence $\{y_n\} \subseteqq Q$ such that

$$\lim_{n\to\infty} \{T(x_n) + y_n\} = b.$$

The limit we consider with respect to the norm in F. Analogous we call a sequence $\{y_n^*\} \subseteqq Q^*$ *admissible* if there exists a sequence $\{x_n^*\} \subseteqq P^*$ such that $\lim_{n\to\infty} \{T^*(y_n^*) - x_n^*\} = a^*$ with respect to the norm in E^*. We denote by $\hat{Z}$ and $\hat{Z}^*$ the sets of admissible sequences $\{x_n\}$ and $\{y_n^*\}$ respectively.

Now we define by

$$\begin{aligned} \omega &= \sup \lim\sup \left\{\langle a^*, x_n \rangle : \{x_n\} \in \hat{Z}\right\}, \\ \omega^* &= \inf \lim\inf \left\{\langle y_n^*, b \rangle : \{y_n^*\} \in \hat{Z}^*\right\} \end{aligned} \tag{3.2.4}$$

the subvalues of the programs. According Definition 3.2.1 a program is subconsistent if $\hat{Z} \neq \emptyset$ and $\hat{Z}^* \neq \emptyset$ respectively.

Lemma 3.2.2. *The set Z is closed and convex.*

Proof: We represent Z in form of $Z = \{x \in P : Tx \in b - Q\}$. The convex cone Q being closed implies that the set $b - Q$ is closed. Because T is a bounded linear operator it follows that the set $\{x \in E : Tx \in b - Q\}$ is closed. Then from P being closed it follows that Z is closed also. Furthermore Z is a intersection of two convex sets thus Z is convex.

Analogously it follows that Z^* is closed and convex. □

Theorem 3.2.1. *A consistent primary program* $[T, P, Q, a^*, b]$ *is subconsistent and we have* $W \leqq \omega$.

Proof: Let $[T, P, Q, a^*, b]$ be consistent and $x_0 \in Z$. It follows $x_0 \in P$ and $b - Tx_0 \in Q$. For all integers n we put $x_n = x_0$ and $y_n = b - Tx_0$. Because

$$\lim_{n\to\infty} \{Tx_n + y_n\} = \lim_{n\to\infty} \{Tx_0 + b - Tx_0\} = b$$

the sequence $\{x_n\} \subseteqq P$ is admissible thus the primary program $[T, P, Q, a^*, b]$ is subconsistent.

Now we consider $W = \sup_Z \langle a^*, x\rangle$. Then for any $\varepsilon > 0$ there exists an element $x_\varepsilon \in Z$ such that

$$W - \varepsilon < \langle a^*, x\rangle. \tag{3.2.5}$$

Obviously the sequence $x_n = x_\varepsilon$ for all n is admissible thus $\{x_n\} \in \hat{Z}$. By (3.2.4), we have

$$\omega \geqq \limsup \langle a^*, x_n\rangle = \langle a^*, x_\varepsilon\rangle$$

and with respect to (3.2.5) we get $\omega \geqq \langle a^*, x_\varepsilon\rangle > W - \varepsilon$. Because this inequality holds for any $\varepsilon > 0$, we have $W \leqq \omega$. □

Similary we can formulate such a theorem for the dual program:

Theorem 3.2.2. *A consistent dual program* $[-T^*, Q^*, P^*, -b, -a^*]$ *is subconsistent and we have* $\omega^* \leqq W^*$.

Moreover we can prove the following theorem:

Theorem 3.2.3. *If the subvalue* ω *of a subconsistent primary program* $[T, P, Q, a^*, b]$ *is finite then there exists an admissible sequence* $\{\bar{x}_n\} \in \hat{Z}$ *such that*

$$\lim_{n\to\infty} \langle a^*, \bar{x}_n\rangle = \omega;$$

if the subvalue ω^* *of a subconsistent dual program* $[-T^*, Q^*, P^*, -b, -a^*]$ *is finite then there exists an admissible sequence* $\{\bar{y}_n{}^*\} \in \hat{Z}^*$ *such that*

$$\lim_{n\to\infty} \langle \bar{y}_n{}^*, b\rangle = \omega^*.$$

We will prove the theorem for the primary program because for the dual program the construction of the proof is analogous. The subvalue ω being finite implies that for any positive ε there exists an admissible sequence $\{x_n{}^\varepsilon\} \in \hat{Z}$ such that

$$\omega - \varepsilon < \limsup \langle a^*, x_n{}^\varepsilon\rangle.$$

According to the definition of admissible sequences there exists a sequence $\{y_n{}^\varepsilon\} \subseteqq Q$ with

$$\lim_{n\to\infty} \{Tx_n{}^\varepsilon + y_n{}^\varepsilon\} = b.$$

Moreover there exists an integer n_ε such that for all $n > n_\varepsilon$ we have $\omega - \varepsilon < \langle a^*, x_n{}^\varepsilon\rangle$. The sequence $\{x_n{}^\varepsilon\}_{n>n_\varepsilon}$ is admissible also because

$$\lim_{\substack{n>n_\varepsilon\\ n\to\infty}} \{Tx_n{}^\varepsilon + y_n{}^\varepsilon\} = b.$$

Then it follows for $n > n_\varepsilon$;

$$\omega - \varepsilon < \langle a^*, x_n^\varepsilon \rangle \leqq \limsup_{n>n_\varepsilon, n\to\infty} \langle a^*, x_n^\varepsilon \rangle = \limsup_{n\to\infty} \langle a^*, x_n^\varepsilon \rangle = \omega$$

thus

$$\lim_{\substack{n>n_\varepsilon \\ n\to\infty}} \langle a^*, x_n^\varepsilon \rangle = \omega .$$

Now we can see that $\{x_n^\varepsilon\}_{n>n_\varepsilon}$ is the required admissible sequence $\{\overline{x}_n\}$.

The next theorem will present the relations concerning the values of the programs.

Theorem 3.2.4. *The consistence of the primary and dual program implies that the values W and W^* are finite and we have $W \leqq W^*$.*

Proof: Because Z and Z^* is not empty we get

$$\inf_{Q^*} \varphi(x, y^*) = \langle a^*, x \rangle \quad \text{for all } x \in Z,$$

$$\sup_{P} \varphi(x, y^*) = \langle y^*, b \rangle \quad \text{for all } y^* \in Z^*.$$

Then we find for all $(x, y^*) \in Z \times Z^*$

$$\langle a^*, x \rangle \leqq \langle y^*, b \rangle$$

which implies $W \leqq W^*$.

The principle of duality for optimization problems (P) and (D) we already could represent by application of the transformation (3.2.3) twice in form of

$$([T, P, Q, a^*, b]^\circledast)^\circledast = [-T^*, Q^*, P^*, -b, -a^*]^\circledast = [T, P, Q, a^*, b].$$

An other characterization of this principle can be given by

$$\begin{aligned}(Z^\circledast)^\circledast &= (\{x \in P : b - Tx \in Q\}^\circledast)^\circledast = \{y^* \in Q^* ; T^*y^* - a^* \in P^*\}^\circledast \\ &= \{x \in P : b - Tx \in Q\} = Z .\end{aligned}$$

Applying the transformation (3.2.3) on the values of the programs we get:

$$\begin{aligned}
W^\circledast &= \left(\sup_Z \langle a^*, x \rangle\right)^\circledast = \sup_{Z^*} \langle y^*, -b \rangle = -\inf_{Z^*} \langle y^*, b \rangle = -W^*, \\
(W^\circledast)^\circledast &= (-W^*)^\circledast = \left(-\inf_{Z^*} \langle y^*, b \rangle\right)^\circledast = -\inf_Z \langle -a^*, x \rangle = \sup_Z \langle a^*, x \rangle = W, \\
\omega^\circledast &= \left(\sup_{\hat{Z}} \limsup_{n\to\infty} \langle a^*, x_n \rangle\right)^\circledast = \sup_{\hat{Z}^*} \limsup_{n\to\infty} \langle y_n^*, -b \rangle \\
&= -\inf_{\hat{Z}^*} \liminf_{n\to\infty} \langle y_n^*, b \rangle = -\omega^*, \\
(\omega^\circledast)^\circledast &= \left(-\inf_{\hat{Z}^*} \liminf_{n\to\infty} \langle y_n^*, b \rangle\right)^\circledast = -\inf_{\hat{Z}} \liminf_{n\to\infty} \langle -a^*, x_n \rangle \\
&= \sup_{\hat{Z}} \limsup_{n\to\infty} \langle a^*, x_n \rangle = \omega .
\end{aligned}$$

The above given characterization of the principle of duality means, it is not essential whether we define problem (P) as a primary program or we consider problem (D) as primary. Depending on the definition of the primary problem always the other one will be the corresponding dual problem. This is an important matter because if we prove any theorem for a primary problem we can formulate a dual theorem by application of transformation (3.2.3) immediately without to make any further demonstrations.

Now we will lead to the main theorem of theory of duality for linear programming known under Duffin theorem [8]. For it we first prove

Lemma 3.2.3. *Consider the primary program* $[T, P, Q, a^*, b]$.
Then the following statements are equivalent:

(i) *For* $x \in Z$ *and* $\lambda \in \mathbb{R}$ *the inequality* $\langle a^*, x\rangle \leqq \lambda$ *holds.*
(ii) *For* $x \in P$, $\lambda \in \mathbb{R}$, $\mu \in \mathbb{R}^+$ *and* $\mu b - Tx \in Q$ *the inequality* $\langle a^*, x\rangle \leqq \mu\lambda$ *holds.*

Proof: Obviously for $\mu = 1$ both statements are the same. Consequently (ii) implies (i). Now we suppose (i). We consider $x \in P$ and $\mu \in \mathbb{R}^+$. First we assume $\mu \neq 0$. Then it follows $x_1 = \dfrac{x}{\mu} \in P$. Moreover $\mu b - Tx \in Q$ implies $b - Tx_1 \in Q$ thus $x_1 \in Z$. According to statement (i) for $\lambda \in \mathbb{R}$ we have $\langle a^*, x_1\rangle \leqq \lambda$. Then it follows immediately $\langle a^*, x\rangle \leqq \mu\lambda$ and this constitute statement (ii). To complete the proof we have to consider the case $\mu = 0$. We have to show that statement (i) implies:

For $x \in P$ and $-Tx \in Q$ the inequality $\langle a^*, x\rangle \leqq 0$ holds. Let us assume the contrary. Then there exists an element $x_1 \in P$ with $-Tx_1 \in Q$ and $\langle a^*, x_1\rangle = \varrho > 0$. Now it follows for all integers $n \geqq 1$, $nx_1 \in P$, $-T(nx_1) \in Q$ and $\langle a^*, nx_1\rangle = n\varrho > 0$. Furthermore $x \in Z$ implies $x + nx_1 \in Z$ and according to statement (i) we have

$$\langle a^*, x\rangle + n\varrho = \langle a^*, x + nx_1\rangle \leqq \lambda$$

for all $n \geqq 1$, but this is not possible. This proves the lemma. □

The investigations in Sections 2.4 and 3.1 show us that for linear programming in finite dimensional spaces the ideas of subconsistence and subvalues of a program is senseless because values and subvalues are equal and consistence and subconsistence is identical. Moreover in the finite dimensional case the programs are always realizable. The situation is quite different for linear programming in abstract Banach spaces. Here the programs could not be realizable and it is quite possible that values and subvalues are different from each other. A theorem of duality for linear programming in finite dimensional spaces says that from consistence of one program it follows the consistence of the other and the values of primary and dual programs are equal. A generalization of this theorem in case of abstract Banach spaces we formulate in form of

Theorem 3.2.5. *The primary program* $[T, P, Q, a^*, b]$ *is consistent and has a finite value* W *then and only then if the dual program* $[-T^*, Q^*, P^*, -b, -a^*]$ *is subconsistent and has a finite subvalue* ω^*. *It is* $W = \omega.^*$

To prove the theorem we first make some preliminary arrangements. Let us consider the space $X = (E^* \times \mathbb{R}) \times (E \times \mathbb{R})$. For elements $(x^*, \alpha) \in E^* \times \mathbb{R}$ and $(x, \xi) \in E \times \mathbb{R}$ we define a bilinear functional in form of

$$\langle (x^*, \alpha), (x, \xi)\rangle = \langle x^*, x\rangle + \alpha\xi .$$

Furthermore let $A: E \times \mathbb{R} \to F$ be a linear mapping in form of $A(x, \xi) = \xi b - Tx$. Then the dual operator A^* characterized by (3.2.1) is a linear mapping from F^* into $E^* \times \mathbb{R}$. The representation of A^* we get from

$$\langle A(x, \xi), y^*\rangle = \langle (x, \xi), A^*y^*\rangle = \langle \xi b - Tx, y^*\rangle = \langle (-T^*y^*, \langle y^*, b\rangle), (x, \xi)\rangle .$$

Consequently

$$A^*y^* = (-T^*y^*, \langle y^*, b\rangle).$$

Since T is a linear bounded mapping the operator A is bounded. Then $(P \times \mathbb{R}^+) \cap A^{-1}(Q) = K$ is a closed convex cone. One may find the representation of the dual convex cone in form of

$$K^* = \overline{(P^* \times \mathbb{R}^+) + A^*(Q^*)}.$$

The proof of this theorem consists of fife steps where the transition from one step to an other is an equivalent transformation.

(i) The primary program $[T, P, Q, a^*, b]$ is consistent and has a finite value W is equivalent to

a$_1$) $\langle a^*, x\rangle \leqq W$ for all $x \in Z$,
b$_1$) for any $\varepsilon > 0$ there exists $x_\varepsilon \in Z$ such that $\langle a^*, x_\varepsilon\rangle > W - \varepsilon$.

According to Lemma 3.2.3 the following two statements are equivalent:

a$_1$) $\langle a^*, x\rangle \leqq W$ for all $x \in Z$,
a$_1$') for $(x, \xi) \in P \times \mathbb{R}^+$ and $A(x, \xi) \in Q$ the inequality $\langle a^*, x\rangle \leqq \xi W$ holds,

(ii) consequently a$_1$) and b$_1$) are equivalent to

a$_2$) for $(x, \xi) \in K$ the inequality $\langle (x, \xi), (-a^*, W)\rangle \geqq 0$ holds,
b$_2$) for any $\varepsilon > 0$ there exists $(x_\varepsilon, 1) \in K$ such that

$$\langle (x_\varepsilon, 1), (-a^*, W - \varepsilon)\rangle < 0$$

(iii) a$_2$) and b$_2$) are equivalent to
a$_3$) $(-a^*, W) \in K^*$
b$_3$) $(-a^*, W - \varepsilon) \notin K^*$

(iv) a$_3$) and b$_3$) are equivalent to
a$_4$) there exist $\{x_n^*\} \subseteqq P^*$, $\{y_n^*\} \subseteqq Q^*$, $\{\alpha_n\} \subseteqq \mathbb{R}^+$ such that

$$\lim_{n\to\infty} \{(x_n^*, \alpha_n) + (-T^*y^*, \langle y_n^*, b\rangle)\}$$
$$= \lim_{n\to\infty} (x_n^* - T^*y_n^*, \alpha_n + \langle y_n^*, b\rangle) = (-a^*, W)$$

i.e.

$$\lim_{n\to\infty} (T^*y_n^* - x_n^*) = a^* \quad \text{and} \quad \lim_{n\to\infty} (\alpha_n + \langle y_n^*, b\rangle) = W.$$

Now it follows that $\{y_n^*\}$ is an admissible sequence thus $\{y_n^*\} \in \hat{Z}^*$ and the dual program $[-T^*, Q^*, P^*, -b, -a^*]$ is subconsistent.

b$_4$) for any $\{y_n^*\} \in \hat{Z}^*$ and for any $\{\alpha_n\} \in \mathbb{R}^+$ the sequence $\{\alpha_n + \langle y_n^*, b\rangle P\}$ does not converge to $W - \varepsilon$, i.e. for any $\{y_n^*\} \in \hat{Z}^*$ we have

$$\liminf_{n\to\infty} \langle y_n^*, b\rangle > W - \varepsilon \quad \text{for } \varepsilon > 0.$$

(v) At last a$_4$) and b$_4$) are equivalent to

a$_5$) for any $\varepsilon > 0$ there exists n_ε such that for $n > n_\varepsilon$ we have

$$W - \varepsilon < \{\alpha_n + \langle y_n^*, b\rangle\} < W + \varepsilon$$

thus $\liminf_{n\to\infty} \langle y_n^*, b\rangle < W + \varepsilon$

b_5) because b_4) holds for any $\varepsilon > 0$ we get

$$\liminf_{n\to\infty} \langle y_n^*, b\rangle \geqq W \quad \text{for } \{y_n^*\} \in \hat{Z}^*.$$

Finally a_5) and b_5) are equivalent to

$$\inf_{\{y_n^*\}\in\hat{Z}^*} \liminf_{n\to\infty} \langle y_n^*, b\rangle = \omega^* = W$$

and this proves the theorem. □

According to the principle of duality we can transform Theorem 3.2.5 to

Corollary 3.2.1. *The dual program* $[-T^*, Q^*, P^*, -b, -a^*]$ *is consistent and has a finite value* W^* *then and only then if the primary program* $[T, P, Q, a^*, b]$ *is subconsistent and has a finite subvalue* $\omega = W^*$.

Moreover we have

Corollary 3.2.2. *From Theorem 3.2.5 and the dual form of the theorem it follows obviously* $W = \omega$ *then and only then if* $W^* = \omega^*$.

On account of Theorem 3.2.1 and Theorem 3.2.5 we can summarize:

If the primary program $[T, P, Q, a^*, b]$ *is consistent and* W *finite then the subvalue* ω *is either finite or* $\omega = \infty$.

In case ω is finite then by Corollary 3.2.1 the dual program $[-T^*, Q^*, P^*, -b, -a^*]$ is consistent and $\omega = W^*$. In case $\omega = \infty$ would never occur then from the consistence of the primary program and a finite value W it follows the consistence of the dual program and the finiteness of the value W^* and reverse. Thus Theorem 3.2.5 makes no sense. Later on we will give an example for the case $\omega = \infty$.

It is obvious that the investigations made in Section 3.1 we can repeat for linear programming in paired spaces. Then Theorem 3.1.1 we can formulate in the following form:

Theorem 3.2.6. *The Lagrangian form*

$$\psi(x, y^*) = -\langle y^*, Tx\rangle + \langle a^*, x\rangle + \langle y^*, b\rangle$$

defined on a convex cone $P \times Q^*$ *produces a pair of dual problems:*

$$\textit{find } W = \max_Z \langle a^*, x\rangle, \tag{P}$$

$$\textit{find } W^* = \min_{Z^*} \langle y^*, b\rangle \tag{D}$$

with

$$Z = \{x \in P : b - Tx \in Q\},$$

$$Z^* = \{y^* \in Q^* : T^*y^* - a^* \in P^*\}.$$

For this problems the following three statements are equivalent:

(i) $\psi(x, y_0^*) \leqq \psi(x_0, y_0^*) \leqq \psi(x_0, y^*)$
for any $(x, y^*) \in P \times Q^*$;

(ii) (P) *and* (D) *are realizable and we have*

$$W = \langle a^*, x_0\rangle = \max_Z \langle a^*, x\rangle = \min_{Z^*} \langle y^*, b\rangle = \langle y_0^*, b\rangle = W^*.$$

(iii) *there exists a pair* $(x_0, y_0^*) \in Z \times Z^*$ *such that*

$$\langle y_0^*, b - Tx_0\rangle = 0, \quad \langle T^*y_0^* - a^*, x_0\rangle = 0.$$

Corollary 3.2.3. *In case* $(x_0, y_0^*) \in Z \times Z^*$ *and* $\langle a^*, x_0\rangle = \langle y_0^*, b\rangle$ *then* x_0 *and* y_0^* *are optimal programs.*

Proof: Obviously we have $\langle a^*, x_0\rangle \leqq \langle T^*y_0^*, x_0\rangle = \langle y_0^*, Tx_0\rangle \leqq \langle y_0^*, b\rangle$. From $\langle a^*, x_0\rangle = \langle y_0^*, b\rangle$ it follows $\langle T^*y_0^* - a^*, x_0\rangle = 0$ and $\langle y_0^*, b - Tx_0\rangle = 0$. Then according to Theorem 3.2.6, (x_0, y_0^*) are optimal programs. □

Furthermore one can easily see that from $\langle a^*, x_0\rangle \leqq W \leqq W^*$ and $\langle y_0^*, b\rangle \geqq W^* \geqq W$ for $(x_0, y_0^*) \in Z \times Z^*$ it follows:

(i) *Let* $[T, P, Q, a^*, b]$ *be consistent and let* W *be the finite value of the primary program. Then for* $y_0^* \in Z^*$ *with the property* $\langle y_0^*, b\rangle = W$ *we have* $\langle y_0^*, b\rangle = W^*$.

(ii) *Let* $[-T^*, Q^*, P^*, -b, -a^*]$ *be consistent and let* W^* *be the finite value of the dual program. Then for* $x_0 \in Z$ *with the property* $\langle a^*, x_0\rangle = W^*$ *we have* $\langle a^*, x_0\rangle = W$.

A criterion for realizability of a program we now present in the following theorem:

Theorem 3.2.7. *Let* $[T, P, Q, a^*, b]$ *be consistent and let* W *be the finite value of the primary program. Then the dual program is realizable and the dual value* W^* *is equal to* W *if the set*

$$\tilde{K}^* = \{(-T^*y^* + x^*, \alpha + \langle y^*, b\rangle) : (x^*, \alpha) \in P^* \times \mathbb{R}^+, y^* \in Q^*\}$$

is closed thus $\tilde{K}^* = K^*$.

Proof: The supposition of Theorem 3.2.7 is equivalent with the statements a_3) and b_3) one can find in the proof of Theorem 3.2.5 thus $(-a^*, W) \in K^*$ and $(-a^*, W - \varepsilon) \notin K^*$ for any $\varepsilon > 0$. $K^* = \tilde{K}^*$ implies that there exist $\bar{x}^* \in P^*$, $\bar{y}^* \in Q^*$, $\bar{\alpha} \in \mathbb{R}^+$ such that $-T^*\bar{y}^* + \bar{x}^* = -a^*$ and $\alpha + \langle \bar{y}^*, b\rangle = W$. Consequently $\bar{y}^* \in Z^*$. Because for any $\varepsilon > 0$ it holds $W - \varepsilon < W \leqq W^*$ and $\bar{y}^*$ being an element in Z^* we get $\langle \bar{y}^*, b\rangle > W - \varepsilon$. The equation $W = \alpha + \langle \bar{y}^*, b\rangle$ now implies $W - \alpha > W - \varepsilon$ for any $\varepsilon > 0$ thus $\alpha = 0$. Then we have $W = \langle \bar{y}^*, b\rangle$ and following statement (i) given above we have $\langle \bar{y}^*, b\rangle = W^*$. This proves the theorem. □

Analogously we can formulate the dual theorem:

Theorem 3.2.8. *Let* $[-T^*, Q^*, P^*, -b, -a^*]$ *be consistent and let* W^* *be the finite value of the dual program. Then the primary program is realizable and the primary value* W *is equal to* W^* *if the set*

$$\tilde{K} = \{(Tx + y, \eta - \langle a^*, x\rangle) : (y, \eta) \in Q \times \mathbb{R}^+, x \in P\}$$

is closed.

Here $\tilde{K}$ is the dual transformation of $\tilde{K}^*$ according to 3.2.3. From the theory of linear programming in finite dimensional spaces we know that a program is realizable and the optimal program is a boundary point of the admissible set at all times. Opposite to the theory in finite dimensional spaces it is possible that a program might not be realizable. But the following theorem holds:

Theorem 3.2.9. *If* $x_0 \in Z$ *is an optimal program of a realizable primary problem* $[T, P, Q, a^*, b]$ *with* $x_0 \neq 0$ *then* x_0 *is a boundary point.*

Proof: Let us assume that x_0 is an interior point of Z. Then there exists $\varepsilon_0 > 0$ such that $S_{\varepsilon_0}(x_0) = \{x \in E: \|x - x_0\| < \varepsilon_0\} \subseteqq Z$. Now we choose ε_1 with $0 < \varepsilon_1 < \varepsilon_0$ and construct

$$x_1 = x_0 + \frac{\varepsilon_1 x_0}{\|x_0\|}, \qquad x_2 = x_0 - \frac{\varepsilon_1 x_0}{\|x_0\|}.$$

The points x_1, x_2 are obviously contained in $S_{\varepsilon_0}(x_0) \subseteqq Z$. Moreover we have in case $\left\langle a^*, \frac{\varepsilon_1 x_0}{\|x_0\|} \right\rangle < 0$

$$\langle a^*, x_2 \rangle = \langle a^*, x_0 \rangle - \left\langle a^*, \frac{\varepsilon_1 x_0}{\|x_0\|} \right\rangle > \langle a^*, x_0 \rangle$$

and in case $\left\langle a^*, \frac{\varepsilon_1 x_0}{\|x_0\|} \right\rangle > 0$

$$\langle a^*, x_1 \rangle = \langle a^*, x_0 \rangle + \left\langle a^*, \frac{\varepsilon_1 x_0}{\|x_0\|} \right\rangle > \langle a^*, x_0 \rangle.$$

This contradicts the fact that x_0 is an optimal program. Thus x_0 must be a boundary point. □

Now we give the dual formulation of Theorem 3.2.9.

Theorem 3.2.10. *If $y_0^* \in Z^*$ is an optimal program of a realizable dual problem $[T^*, Q^*, P^*, -b, -a^*]$ with $y_0^* \neq 0$ then y_0^* is a boundary point of Z^*.*

Now we consider the special case $a^* \equiv 0$ or $b \equiv 0$ respectively. A primary program $[T, P, Q, 0, b]$ or $[T, P, Q, a^*, 0]$ is said to be homogenous. The dual transformed homogenized programs are:

$$[T, P, Q, 0, b]^{\circledast} = [-T^*, Q^*, P^*, -b, 0],$$
$$[T, P, Q, a^*, 0]^{\circledast} = [-T^*, Q^*, P^*, 0, -a^*].$$

Because here only cones P, Q with vertexes at the origin are considered the homogenized programs $[T, P, Q, a^*, 0]$, $[-T^*, Q^*, P^*, -b, 0]$ are consistent and subconsistent. The value or subvalue of a homogenized program is either 0 or $+\infty$.

A consequence of Theorem 3.2.5 and Corollary 3.2.1 are the following statements:

Corollary 3.2.4. *The primary program $[T, P, Q, a^*, b]$ is consistent if and only if the homogenized program $[-T^*, Q^*, P^*, -b, 0]$ has the subvalue zero.*

The dual program $[-T^*, Q^*, P^*, -b, -a^*]$ is consistent if and only if the homogenized program $[T, P, Q, a^*, 0]$ has the subvalue zero.

Proof: Obviously the consistence of $[T, P, Q, a^*, b]$ is equivalent to the consistence of the homogeneous program $[T, P, Q, 0, b]$ and the value of the homogeneous program is zero. Then from Theorem 3.2.5 it follows the equivalence to the subconsistence of the dual transformed program $[T, P, Q, 0, b]^{\circledast} = [-T^*, Q^*, P^*, -b, 0]$ which has a subvalue zero. □

Analogously we have

Corollary 3.2.5. *The primary program $[T, P, Q, a^*, b]$ is subconsistent if and only if the homogenized dual program $[-T^*, Q^*, P^*, -b, 0]$ has the value zero. The dual program $[-T^*, Q^*, P^*, -b, -a^*]$ is subconsistent if and only if the homogenized primary program $[T, P, Q, a^*, 0]$ has the value zero.*

Definition 3.2.2. We call a primary program *superconsistent* if $Z \neq 0$ and if there exist an $x \in Z$ such that $b - Tx$ is an interior point of Q; analogously the dual program is said to be *superconsistent* if $Z^* \neq 0$ and if there exists an $y^* \in Z^*$ such that $T^*y^* - a^*$ is an interior point of P^*.

Theorem 3.2.1, 3.2.2, 3.2.5 describe the correlations between consistent and subconsistent programs. Now we will investigate the correlation between consistent and superconsistent programs. Obviously a superconsistent program is consistent and subconsistent. If one program is consistent by Theorem 3.2.5 we know that the value of the program is equal to the subvalue of the other one. From this we should guess in case one program is superconsistent that the values of the primary and dual program are equal.

Theorem 3.2.11. *The dual program* $[-T^*, Q^*, P^*, -b, -a^*]$ *is consistent and realizable if the primary program* $[T, P, Q, a^*, b]$ *is superconsistent and has a finite value* W.

The primary program is consistent and realizable if the dual program is superconsistent and has a finite value W^*.

Proof: Let $[T, P, Q, a^*, b]$ be superconsistent and let W be its finite value. We will show that there exists an element $y^* \in Z^*$ such that $\langle y^*, b\rangle \leqq W$. Analogous Lemma 3.2.3, one can prove that the following two statements are equivalent:

(i) For $y^* \in Z^*$ and $\lambda \in \mathbb{R}$ the inequality $\langle y^*, b\rangle \leqq \lambda$ holds.

(ii) For $y^* \in Q^*$, $\lambda \in \mathbb{R}$, $\mu \in \mathbb{R}^+$ and $T^*y^* - \mu a^* \in P^*$ the inequality $\langle y^*, b\rangle \leqq \mu\lambda$ holds.

Consequently $y^* \in Z^*$ imply $\langle y^*, b\rangle \leqq W$ is equivalent to

$$y^* \in Q^*,\ \mu \in \mathbb{R}^+ \text{ and } T^*y^* - \mu a^* \in P^* \text{ imply } \langle y^*, b\rangle \leqq \mu W. \tag{3.2.6}$$

Since $y^* \in Q^*$ is equivalent to $\langle y^*, y\rangle \geqq 0$ for all $y \in Q$, $T^*y^* - \mu a^* \in P^*$ is equivalent to $\langle T^*y^* - \mu a^*, x\rangle \geqq 0$ for all $x \in P$. $\mu W - \langle y^*, b\rangle \geqq 0$ is equivalent to $t(\mu W - \langle y^*, b\rangle) \geqq 0$ for all $t \in \mathbb{R}^+$. We can express (3.2.6) in terms of functionals, i.e. for all $x \in P$, $y \in Q$, $t \in \mathbb{R}^+$ we have

$$\langle T^*y^* - a^*, x\rangle + \langle y^*, y\rangle + t(\mu W - \langle y^*, b\rangle) \geqq 0.$$

Now we assume that there does not exist an element $y^* \in Z^*$ with $\langle y^*, b\rangle \leqq 0$. Then the statement: for all $x \in P$, $y \in Q$, $t \in \mathbb{R}^+$ we have

$$\langle T^*y^* - a^*, x\rangle + \langle y^*, y\rangle + t(\mu W - \langle x^*, b\rangle) \geqq 0$$

implies $-\mu \geqq 0$. It follows for all $x \in P$, $y \in Q$, $t \in \mathbb{R}^+$

$$\langle y^*, Tx + y - tb\rangle + \mu[tW - \langle a^*, x\rangle] \geqq 0 \text{ implies } -\mu \geqq 0. \tag{3.2.7}$$

We can express (3.2.7) by the statement: for all $(x, y, t) \in P \times Q \times \mathbb{R}^+$ it holds

$$\langle (y^*, \mu), (Tx + y - tb, tW - \langle a^*, x\rangle)\rangle \geqq 0 \quad \text{implies } -\mu \geqq 0.$$

Obviously the set $K = \{(Tx + y - tb, tW - \langle a^*, x\rangle) \colon (x, y, t) \in P \times Q \times \mathbb{R}^+\}$ is a convex cone in $F \times \mathbb{R}$. According to (3.2.7) we get the dual cone K^* in form of $K^* = Q^* \times \mathbb{R}^-$. Since $\langle (y^*, \alpha), (0, -1)\rangle = -\alpha \geqq 0$ for all $(y^*, \alpha) \in K^*$ by Lemma 3.2.1 the element $(0, -1)$ is contained in the closure $\overline{K}$. Thus there exists a sequence $\{(x_n, y_n, t_n)\} \subseteqq P \times Q \times \mathbb{R}^+$ such that

$$\lim_{n\to\infty} (Tx_n + y_n - t_n b) = 0, \quad \lim_{n\to\infty} (t_n W - \langle a^*, x_n\rangle) = -1.$$

Now we consider any $\varepsilon > 0$. Since

$$W = \sup_{x \in Z} \langle a^*, x\rangle$$

is finite there exists an element $x_\varepsilon \in Z$ such that $\langle a^*, x_\varepsilon\rangle > W - \varepsilon$. According to the assumption we choose an interior point $x_0 \in Z$ and put $\bar{x} = (1 - \delta)\, x_0 + \delta x$. For a suitably small positive number δ with $0 < \delta < 1$ the element x is an interior point of Z. Furthermore for a sufficiently small δ it is

$$\langle a^*, \bar{x}\rangle = \langle a^*, x_\varepsilon\rangle + \delta\langle a^*, x_0 - x_\varepsilon\rangle > W - 2\varepsilon. \tag{3.2.8}$$

Now for the sequence $\bar{x}_n = x_n + \bar{x}$ it follows

$$\lim_{n\to\infty} \big(T\bar{x}_n + y_n - b(1 + t_n)\big) = T\bar{x} - b \notin Q \tag{3.2.9}$$

and with respect to (3.2.8) we get

$$\lim_{n\to\infty} \big(\langle a^*, \bar{x}_n\rangle - W(1 + t_n)\big) = 1 + \langle a^*, \bar{x}\rangle - W > 1 - 2\varepsilon. \tag{3.2.10}$$

Thus for sufficiently large n we get from (3.2.10)

$$-\varepsilon < \langle a^*, \bar{x}_n\rangle - W(1 + t_n) - 1 - \langle a^*, \bar{x}\rangle + W < \varepsilon$$

and consequently

$$\langle a^*, \bar{x}_n\rangle > W(1 + t_n) + 1 + \langle a^*, \bar{x}\rangle - W - \varepsilon > W(1 + t_n) + 1 - 3\varepsilon.$$

The statement (3.2.9) implies $b(1 + t_n) - T\bar{x}_n$ being an interior point of Q for large n thus $\bar{\bar{x}} = \dfrac{x_n}{1 + t_n}$ is an interior point of Z. Moreover if we put $\varepsilon = \dfrac{1}{4}$ we get

$$\langle a^*, \bar{\bar{x}}\rangle = \frac{1}{1 + t_n}\langle a^*, \bar{x}_n\rangle > W + \frac{1}{1 + t_n} - \frac{3}{4(1 + t_n)} = W + \frac{1}{4(1 + t_n)} > W$$

and this contradicts the definition of W. Thus there exists an element $y^* \in Z^*$ such that $\langle y^*, b\rangle \leqq W$. Then the dual program is consistent which is the assertion of the theorem. □

Finally we present conditions for a program to be superconsistent. We assume that the convex cones Q and P^* have interior points.

Theorem 3.2.12. *The primary program* $[T, P, Q, a^*, b]$ *is superconsistent if and only if the homogenized dual program has the value zero and this value is attained only by* $y_0{}^* = 0$.

The dual program $[-T^*, Q^*, P^*, -b, -a^*]$ *is superconsistent if and only if the homogenized primary program has the value zero and this value is attained only by* $x_0 = 0$.

Proof: The proof of this theorem consists of three steps.

(i) If y is an interior point of Q and $y^* \in Q^*$ with $y^* \neq 0$ then it holds $\langle y^*, y\rangle > 0$.

To prove (i) we assume $\langle y^*, y\rangle = 0$. There exists an element $\bar{y} \in F$ such that $\langle y^*, \bar{y}\rangle < 0$. Since y is an interior point for sufficiently small $\varepsilon > 0$ the element $\bar{\bar{y}} = y + \varepsilon\bar{y}$ is contained in Q. But $\langle y^*, \bar{\bar{y}}\rangle = \varepsilon\langle y^*, \bar{y}\rangle < 0$ and this is not possible.

(ii) We suppose the primary program to be superconsistent i.e. there exists an interior point $x \in Z$. Now it follows that $b - Tx$ is an interior point of Q and according to (i) we get for any $y^* \in Q^*$ with $y^* \neq 0$ the inequality $\langle y^*, b - Tx\rangle = \langle y^*, b\rangle - \langle T^*y^*, x\rangle > 0$. This inequality is valid specially for admissible points of the homogenized dual program, i.e. $y^* \in Z_0{}^* = \{y^* \in Q^*\colon T^*y^* \in P^*\}$ and $y^* \neq 0$. Since $T^*y^* \in P^*$ finally we get for any $y^* \in Z_0{}^*$ with $y^* \neq 0$

$$\langle y^*, b\rangle > \langle T^*y^*, x\rangle \geqq 0$$

thus $\inf\limits_{Z_0{}^*} \langle y^*, b\rangle = \langle y_0{}^*, b\rangle = 0$ with $y_0{}^* = 0$.

(iii) Now we suppose that the homogenized program has the value zero and this value is attained only by $y_0^* = 0$, i.e. for any $y^* \in Z_0^*$ we have $\langle y^*, b\rangle \geqq 0$ and

$$\inf_{Z_0^*} \langle y^*, b\rangle = \langle y_0^*, b\rangle = 0 .$$

Since Q has interior points it follows that the convex cone $T(P) + Q$ has interior points. We point out that b is an interior point of $T(P) + Q$.

Let us assume the reverse. By separation Theorem 1.2.4 there exists an element $\bar{y}^* \in F^*$ and $\bar{y}^* \neq 0$ such that for all $(x, y) \in P \times Q$ we have

$$\langle \bar{y}^*, Tx + y\rangle \geqq 0, \quad \langle y^*, b\rangle \leqq 0 . \tag{3.2.11}$$

Since only cones with vertex at the origin are considered it follows from (3.2.11) for $x = 0$ and $y = 0$ respectively

$$\langle \bar{y}^*, y\rangle \geqq 0 \quad \text{for all } y \in Q, \text{ thus } \bar{y}^* \in Q^*,$$

$$\langle \bar{y}^*, Tx\rangle = \langle T^*\bar{y}^*, x\rangle \geqq 0 \quad \text{for all } x \in P, \text{ thus } T^*\bar{y}^* \in P^* .$$

Consequently $\bar{y}^* \in Z_0^*$ and $\bar{y}^* \neq 0$. According to the superposition of (iii) we get $\langle \bar{y}^*, b\rangle > 0$ and this contradicts (3.2.11) hence b is an interior point of $T(P) + Q$.

Now we choose an interior point $y_0 \in Q$. For $\varepsilon > 0$ it is $\bar{y} = \varepsilon y_0 \in Q$ interior point. Since b is an interior point of $T(P) + Q$ in case ε sufficiently small the point $b - \varepsilon y_0$ will be an interior point of $T(P) + Q$. Then there exists a sequence $(x_n, y_n) \in P \times Q$ such that $\lim\limits_{n\to\infty} (Tx_n + y_n) = b - y_0 = b - \frac{1}{\varepsilon}\bar{y}$. We put $y_n^0 = b - \bar{y} - Tx_n - y_n$. Obviously $\lim\limits_{n\to\infty} y_n^0 = 0$. For large n, $\bar{\bar{y}} = \bar{y} + y_n^0$ is an interior point of Q hence $b - Tx_n = \bar{\bar{y}} + y_n$ is an interior point of Q. Consequently the primary program is superconsistent which proves the theorem □.

We will continue our considerations by constructing some examples for linear programming in abstract Banach spaces. For that we consider any closed interval $[0, b]$ in $\mathbb{R}$. We denote

$$\mathbf{L}_2[0, b] = \left\{x(t) : t \in [0, b], \int_0^b [x(t)]^2 \, dt < \infty\right\}.$$

Obviously $\mathbf{L}_2[0, b]$ is a Hilbert space and $\langle x, \bar{x}\rangle = \int_0^b x(t)\, \bar{x}(t)\, dt$ the scalar product of elements x and $\bar{x}$ contained in $\mathbf{L}_2[0, b]$. According to the theorem of F. Riesz the continuous linear functionals defined on $\mathbf{L}_2[0, b]$ can be represented in form of a scalar product $x^*(x) = \int_0^b x^*(t)\, x(t)\, dt$ thus $\mathbf{L}_2^*[0, b] = \mathbf{L}_2[0, b]$.

Example 1. Let $E = \mathbf{L}_2[0, 1] \times \mathbb{R}$ and let $F = \mathbf{L}_2[0, 1]$. Then the dual spaces we get in form of $E^* = \mathbf{L}_2[0, 1] \times \mathbb{R}$, $F^* = \mathbf{L}_2[0, 1]$. If we denote $\mathbf{L}_2^+[0, 1] = \{x \in \mathbf{L}_2[0, 1];\ x(t) \geqq 0\}$ then we consider the convex cones $P = \mathbf{L}_2^+[0, 1] \times \mathbb{R}^+ \subseteqq E$ and $Q = \mathbf{L}_2^+[0, 1] \subseteqq F$. It is obvious that $P^* = P$ and $Q^* = Q$. Next we define the linear mapping T from E into F by

$$T(x, \xi) = -\int_t^1 x(\tau)\, d\tau - \xi .$$

The corresponding dual operator T^* is defined by the equation

$$\langle y^*, T(x, \xi)\rangle = \int_0^1 y^*(t)\left(-\int_t^1 x(\tau)\,\mathrm{d}\tau - \xi\right)\mathrm{d}t = \langle T^*y^*, (x, \xi)\rangle$$

for all $(x, \xi) \in \mathbf{L}_2[0, 1] \times \mathbb{R}$ and all $y^* \in \mathbf{L}_2[0, 1]$. Now we calculate

$$\begin{aligned}\langle y^*, T(x, \xi)\rangle &= \xi \int_0^1 \left(-y^*(t)\right)\mathrm{d}t + \int_0^1 \left(-y^*(t)\right)\int_t^1 x(\tau)\,\mathrm{d}\tau\,\mathrm{d}t \\ &= \xi \int_0^1 \left(-y^*(t)\right)\mathrm{d}t - \left(\left[-\int_1^t x(\tau)\,\mathrm{d}\tau \int_0^t y^*(\tau)\,\mathrm{d}\tau\right]_0^1 \right.\\ &\quad \left. - \int_0^t \left(-x(t)\right)\int_0^t y^*(\tau)\,\mathrm{d}\tau\,\mathrm{d}t\right) \\ &= \int_0^1 x(t)\int_0^t \left(-y^*(\tau)\right)\mathrm{d}\tau\,\mathrm{d}t + \xi\int_0^1 \left(-y^*(t)\right)\mathrm{d}t \\ &= \left\langle (x, \xi), \left(\int_0^t \left(-y^*(\tau)\right)\mathrm{d}\tau, \int_0^1 \left(-y^*(t)\right)\mathrm{d}t\right)\right\rangle\end{aligned}$$

thus

$$T^*y^* = \left(-\int_0^t y^*(\tau)\,\mathrm{d}\tau, -\int_0^1 y^*(t)\,\mathrm{d}t\right).$$

Now we consider the Lagrangian form for $b(t) = -1$, $0 \leqq t \leqq 1$, and $a^* = (-t, -2) \in E^* = \mathbf{L}_2[0, 1] \times \mathbb{R}$ defined on $P \times Q^* = (\mathbf{L}_2^+ \times \mathbb{R}^+) \times \mathbf{L}_2^+$: $\psi(x, y^*) = -\langle y^*, T(x, \xi)\rangle + \langle(-t, -2), (x, \xi)\rangle + \langle y^*, -1\rangle = -\int_0^1 tx(t)\,\mathrm{d}t - 2\xi + \int_0^1 y^*(t)\left(\int_t^1 x(\tau)\,\mathrm{d}\tau + \xi - 1\right)\mathrm{d}t.$

The primary program will be produced by

$$\begin{aligned}\inf_{Q^*} \psi(x, y^*) &= -\int_0^1 tx(t)\,\mathrm{d}t - 2\xi + \inf_{Q^*} \int_0^1 y^*(t)\left(\int_t^1 x(\tau)\,\mathrm{d}\tau + \xi - 1\right)\mathrm{d}t \\ &= -\int_0^1 tx(t)\,\mathrm{d}t - 2\xi\end{aligned}$$

with $(x, \xi) \in Z = \left\{(x, \xi) \in \mathbf{L}_2^+ \times \mathbb{R}^+ : \int_t^1 x(\tau)\,\mathrm{d}\tau + \xi - 1 \geqq 0, 0 \leqq t \leqq 1\right\}.$

The dual program $[-T^*, Q^*, P^*, -b, -a^*]$ we get from

$$\begin{aligned}\sup_P \psi(x, y^*) &= -\int_0^1 y^*(t)\,\mathrm{d}t + \sup_P \langle (x, \xi), (-t, -2) - T^*y^*\rangle \\ &\quad - \int_0^1 y^*(t)\,\mathrm{d}t + \sup_P \left\langle (x, \xi), \left(\int_0^t y^*(\tau)\,\mathrm{d}\tau - t, \int_0^1 y^*(\tau)\,\mathrm{d}\tau - 2\right)\right\rangle \\ &= -\int_0^1 y^*(t)\,\mathrm{d}t + \sup_P \left(\int_0^1 x(t)\left(\int_0^t y^*(\tau)\,\mathrm{d}\tau - t\right)\mathrm{d}t\right. \\ &\quad \left. + \xi\left(\int_0^1 y^*(\tau)\,\mathrm{d}\tau - 2\right)\right) = -\int_0^1 y^*(t)\,\mathrm{d}t\end{aligned}$$

with

$$y^* \in Z^* = \left\{y^* \in \mathbf{L}_2^+ : \int\limits_0^t y^*(\tau)\,\mathrm{d}\tau - t \leqq 0;\ 0 \leqq t \leqq 1;\ \int\limits_0^1 y^*(\tau)\,\mathrm{d}\tau - 2 \leqq 0\right\}.$$

Now we have the pair of dual problems in form of

$$W = \sup_Z \left(-\int\limits_0^1 tx(t)\,\mathrm{d}t - 2\xi\right), \tag{P}$$

$$W^* = \inf_{Z^*} \left(-\int\limits_0^1 y^*(t)\,\mathrm{d}t\right). \tag{D}$$

Obviously both problems are realizable and the optimal solutions are $(x_0, \xi_0) = (0, 1)$ and $y_0^* = 1$ thus $W = -2$ and $W^* = -1$. Consequently we have constructed an example where the values of the problems are different from each other. Such a case we will call a duality gap which does not appear in the finite dimensional theory. For the homogenized dual program i.e. $a^* = (0, 0)$ we get $Z^* = \{0\}$ thus $W^* = 0$. Hence by Theorem 3.2.12 the primary program must be super-consistent.

According Theorem 3.2.3 there exist admissible sequences $(\bar{x}_n, \bar{\xi}_n) \in \hat{Z}$ and $\bar{y}_n^* \in \hat{Z}^*$ such that

$$\lim_{n\to\infty} \left(-\int\limits_0^1 t\bar{x}_n\,\mathrm{d}t - 2\bar{\xi}_n\right) = \omega = W^* = -1,$$

$$\lim_{n\to\infty} \left(-\int\limits_0^1 \bar{y}_n^*(t)\,\mathrm{d}t\right) = \omega^* = W^* = -2.$$

If we choose $\bar{y}_n(t) = 0$ we find that $(\bar{x}_n, \bar{\xi}_n) = (nt^{n-1}, 0)$ is an admissible sequence and it is

$$\lim_{n\to\infty} \left(T(\bar{x}_n, \bar{\xi}_n) + \bar{y}_n\right) = \lim_{n\to\infty} \left(-\int\limits_t^1 n\tau^{n-1}\,\mathrm{d}\tau\right) = -1.$$

If we choose $(\bar{x}_n^*, \bar{\alpha}_n) \in P^* = \mathbf{L}_2^+ \times \mathbb{R}^+$ in form of $(\bar{x}_n^*, \bar{\alpha}_n) = (0, 0)$ we obviously find that for $\bar{y}_n^* = 1 + nt^{n-1}$ we get

$$\lim_{n\to\infty} \left(-\int\limits_0^t (1 + n\tau^{n-1}\,\mathrm{d}\tau), -\int\limits_0^1 (1 + n\tau^{n-1}\,\mathrm{d}\tau)\right) = (-t, -2)$$

thus $\bar{y}_n^* = 1 + nt^{n-1} \subseteqq \hat{Z}^*$. Furthermore it is

$$\lim_{n\to\infty} \left(-\int\limits_0^1 (1 + nt^{n-1})\,\mathrm{d}t\right) = -2.$$

Example 2. We will construct an example of a homogeneous primary program $[T, P, Q, a^*, 0]$ with a finite value and an infinite subvalue. The spaces E, F and the convex cones P, Q we choose likewise as in Example 1. The linear mapping from E into F we define as

$$T(x, \xi) = \xi - \int\limits_t^1 x(\tau)\,\mathrm{d}\tau.$$

Furthermore let $a^* \in E^* = \mathbf{L}_2[0, 1] \times \mathbb{R}$ be $a^* = (-1, 2)$. Since we assume the primary program to be homogeneous it is $b(t) \equiv 0$. The corresponding Lagrangian form

$$\varphi(x, y^*) = -\langle y^*, T(x, \xi)\rangle + \langle(-1, 2), (x, \xi)\rangle$$

produces the following primary problem:

$$\inf_{Q^*} \varphi(x, y^*) = -\int_0^1 x(t)\, dt + 2\xi + \inf_{Q^*} \langle y^*, -T(x, \xi)\rangle.$$

It is

$$\inf_{Q^*} \langle y^*, -T(x, \xi)\rangle = \inf_{Q^*} \int_0^1 y^*(t) \left(\int_t^1 x(\tau)\, d\tau - \xi\right) dt = 0$$

for all $(x, \xi) \in Z = \left\{(x, \xi) \in \mathbf{L}_2^+ \times \mathbb{R}^+ : \int_t^1 x(\tau)\, d\tau - \xi \geqq 0, 0 \leqq t \leqq 1\right\} = \mathbf{L}_2^+ \times \{0\}$.

The primary problem takes the form

$$W = \sup_Z \left(2\xi - \int_0^1 x(t)\, dt\right). \tag{P}$$

Obviously $(0, 0) \in Z$ consequently $W = 0$.

The dual operator $T^*: \mathbf{L}_2[0, 1] \to \mathbf{L}_2[0, 1] \times \mathbb{R}$ we get analogously as in Example 1 in form of

$$T^* y^* = \left(-\int_0^t y^*(\tau)\, d\tau, \int_0^1 y^*(t)\, dt\right).$$

Then from the Lagrangian form $\varphi(x, y^*)$ it follows for the dual problem:

$$\sup_P \varphi(x, y^*) = \sup_P \langle(x, \xi), (-1, 2) - T^* y^*\rangle$$

$$= \sup_P \left(\int_0^1 x(t) \left(\int_0^t y(\tau)\, d\tau - 1\right) dt + \xi\left(2 - \int_0^1 y^*(t)\, dt\right)\right) = 0$$

for all

$$y^* \in Z^* = \left\{y^* \in \mathbf{L}_2^+ : \int_0^t y(\tau)\, d\tau - 1 \leqq 0, 0 \leqq t \leqq 1; 2 - \int_0^1 y^*(t)\, dt \leqq 0\right\}.$$

The set of admissible points Z^* is obviously empty. But the primary problem is also subconsistent. If we choose

$$(\bar{x}_n, \bar{\xi}_n) = (n^{5/4} t^{n-1}, n^{1/4})$$

and $\bar{y}_n(t) \equiv 0$ then it follows

$$\lim_{n\to\infty} \left(T(\bar{x}_n, \bar{\xi}_n) + \bar{y}_n\right) = \lim_{n\to\infty} \left(n^{1/4} - \int_t^1 n^{5/4} \tau^{n-1}\, d\tau\right) = \lim_{n\to\infty} n^{1/4} t^n.$$

$$= 0, \quad 0 < t < 1.$$

Thus $(n^{5/4} t^{n-1}, n^{1/4}) \in \hat{Z}$ is an admissible sequence and since

$$\lim_{n\to\infty} \langle a^*, (\bar{x}_n, \bar{\xi}_n)\rangle = \lim_{n\to\infty} \left(2n^{1/4} - \int_0^1 n^{5/4} t^{n-1}\, dt\right) = \lim_{n\to\infty} n^{1/4} = \infty$$

the subvalue ω of the primary program is infinitive.

Example 3. Next we define a Lagrangian form such that the primary program has a finite value W but the program is not realizable. Let us consider $E = F = \mathbf{L}_2[0, 1]$ and $P = Q = \mathbf{L}_2^+$. The linear mapping T from $\mathbf{L}_2[0, 1]$ into $\mathbf{L}_2[0, 1]$ we define in form of

$$Tx \equiv -\int_0^1 x(t)\, dt, \quad 0 \leqq t \leqq 1.$$

Furthermore we put $b \equiv -1$ and $a^* = -t^2$. Then the corresponding Lagrangian form defined on $\mathbf{L}_2^+ \times \mathbf{L}_2^+$ appears as

$$\varphi(x, y^*) = \int_0^1 y^*(t) \left(\int_0^1 x(\tau)\, d\tau\right) dt - \int_0^1 t^2 x(t)\, dt - \int_0^1 y^*(t)\, dt.$$

For the primary problem (P) we calculate

$$\inf_{Q^*} \varphi(x, y^*) = -\int_0^1 t^2 x(t)\, dt + \inf_{Q^*} \int_0^1 y^*(t) \left(\int_0^1 x(t)\, dt - 1\right) dt = -\int_0^1 t^2 x(t)\, dt$$

for all $x \in Z = \left\{x \in \mathbf{L}_2^+ : \int_0^1 x(t)\, dt - 1 \geqq 0\right\}$. Obviously $x \equiv 0$ is not contained in Z. But the primary program is consistent because the functions

$$x_n(t) = \begin{cases} n, & 0 \leqq t \leqq 1/n, \\ 0, & 1/n < t \leqq 1, \end{cases}$$

are admissible. This follows immediately from

$$\int_0^1 x_n(t)\, dt = \int_0^{1/n} n\, dt = 1.$$

Now we get

$$W = \sup_Z \left(-\int_0^1 t^2 x(t)\, dt\right) \leqq 0. \tag{P}$$

Because $0 \notin Z$, but $x_n(t) \in Z$ and

$$\lim_{n\to\infty} \left(-\int_0^1 t^2 x_n(t)\, dt\right) = \lim_{n\to\infty} \left(-\int_0^{1/n} t^2 n\, dt\right) = \lim_{n\to\infty} (-1/3n^2) = 0$$

the primary program has the finite value zero and this program is not realizable.

Example 4. We introduce the Hilbert space

$$\begin{aligned} \mathbf{L}_2^n[0, 1] &= \mathbf{L}_2[0, 1] \times \cdots \times \mathbf{L}_2[0, 1] \\ &= \{x(t) = (x_1(t), \ldots, x_n(t)) : x_i \in \mathbf{L}_2[0, 1];\ i = 1, \ldots, n\}. \end{aligned}$$

Obviously it is $\mathbf{L}_2^1[0, 1] = \mathbf{L}_2[0, 1]$. We denote the scalar product in $\mathbf{L}_2^n[0, 1]$ by $\langle \cdot, \cdot \rangle_n$ and we define

$$\langle x, \bar{x} \rangle_{(n)} = \sum_{i=1}^n \langle x_i, \bar{x}_i \rangle_{(1)} = \sum_{i=1}^n \int_0^1 x_i(t)\, \bar{x}_i(t)\, dt, \quad x, \bar{x} \in \mathbf{L}_2^n[0, 1]. \tag{3.2.12}$$

The scalar product $\langle .,.\rangle_{(n)}$ generates a norm in $\mathbf{L}_2{}^n[0, 1]$ by

$$\|x\|_{(n)} = (\langle x, x\rangle_{(n)})^{1/2} = \left(\sum_{i=1}^{n} \int_0^1 x_i(t)^2 \, dt\right)^{1/2}.$$

Now we put $E = \mathbf{L}_2{}^n[0, 1]$ and $F = \mathbf{L}_2{}^{n-1}[0, 1]$. According to the theorem of F. Riesz and the scalar product defined in (3.2.12) we get $E^* = E$ and $F^* = F$. If we denote

$$\mathbf{L}_2{}^{n+} = \{x \in \mathbf{L}_2{}^n[0, 1] \colon x_i(t) \geqq 0,\ i = 1, \ldots, n;\ 0 \leqq t \leqq 1\}$$

then you find out that $\mathbf{L}_2{}^{n+}$ is a convex cone, moreover for the dual convex cone we get $(\mathbf{L}_2{}^{n+})^* = L_2{}^{n+}$. We choose the convex cones $P = \mathbf{L}_2{}^{n+}$ and $Q = \mathbf{L}_2{}^{(n-1)+}$. Obviously it is $P^* = P$ and $Q^* = Q$. Let $A = (a_{ij})$ be a real matrix of order $(n-1, n)$ and let $B = (b_{ij})$ be a real matrix of order $(n-1, n)$. By A^* and B^* we denote the corresponding transposed matrix. We define a linear mapping from E into F in form of

$$Tx = Ax + \int_0^t Bx(\tau) \, d\tau \in \mathbf{L}_2{}^{(n-1)}[0, 1].$$

Here expressions Ax and $\int_0^t Bx(\tau)\, d\tau$ mean:

$$Ax = \begin{pmatrix} a_{11} & \cdots & a_{1,n} \\ a_{n-1,1} & \cdots & a_{n-1,n} \end{pmatrix} \begin{pmatrix} x_1(t) \\ \vdots \\ x_n(t) \end{pmatrix} = \begin{pmatrix} \sum_{i=1}^{n} a_{1,i} x_i(t) \\ \vdots \\ \sum_{i=1}^{n} a_{n-1,i} x_i(t) \end{pmatrix}$$

$$\int_0^t Bx(\tau)\, d\tau = \begin{pmatrix} \sum_{i=1}^{n} b_{1,i} \int_0^t x_i(\tau)\, d\tau \\ \vdots \\ \sum_{i=1}^{n} b_{n-1,i} \int_0^t x_i(\tau)\, d\tau \end{pmatrix}.$$

The dual operator T^* is a linear mapping from $\mathbf{L}_2{}^{n-1}[0, 1]$ into $\mathbf{L}_2{}^n[0, 1]$. To calculate the dual operator T^* we have to consider the equation

$$\langle Tx, y^*\rangle_{(n-1)} = \langle x, T^*y^*\rangle_{(n)}, \quad x \in E = \mathbf{L}_2{}^n[0, 1], \quad y \in F^* = \mathbf{L}_2{}^{n-1}[0, 1].$$

It is

$$\langle Tx, y^*\rangle_{(n-1)} = \left\langle Ax + \int_0^t Bx(\tau)\, d\tau, y^* \right\rangle_{(n-1)}$$

$$= \langle Ax, y^*\rangle_{(n-1)} + \left\langle \int_0^t Bx(\tau)\, d\tau, y^* \right\rangle_{(n-1)}. \tag{3.2.13}$$

Further calculations we have to convey with the aim to separate the element x. The first term leads to

$$\langle Ax, y^*\rangle_{(n-1)} = \sum_{i=1}^{n-1} \int_0^1 y_i^*(t) \sum_{j=1}^{n} a_{ij} x_j(t)\, dt$$

$$= \int_0^1 \sum_{j=1}^{n} x_j(t) \sum_{i=1}^{n-1} a_{ij} y^*(t)\, dt = \langle x, A^*y^*\rangle_{(n)}. \tag{3.2.14}$$

From the second term we get

$$\left\langle \int_0^t Bx(\tau)\,d\tau, y^* \right\rangle_{(n-1)} = \sum_{i=1}^{n-1} \int_0^1 y_i^*(t) \left(\sum_{j=1}^{t} b_{ij} \int_0^t x_j(\tau)\,d\tau \right) dt$$

$$= \int_0^1 \sum_{j=1}^{n} \int_0^t x_j(\tau)\,d\tau \sum_{i=1}^{n-1} b_{ij} y_i^*(t)\,dt.$$

By application of partial integration for a fixed j it follows

$$\int_0^1 \left(\int_0^t x_j(\tau)\,d\tau \right) \left(\sum_{i=1}^{n-1} b_{ij} y_i^*(t) \right) dt$$

$$= \left[\int_0^t x_j(\tau)\,d\tau \int_0^t \sum_{i=1}^{n-1} b_{ij} y_i^*(\tau)\,d\tau \right]_0^1 - \int_0^1 x_j(t) \left(\int_0^t \sum_{i=1}^{n-1} b_{ij} y_i^*(\tau)\,d\tau \right) dt$$

$$= \int_0^1 x_j(\tau)\,d\tau \int_0^1 \sum_{i=1}^{n-1} b_{ij} y_i^*(\tau)\,d\tau - \int_0^1 x_j(t)$$

$$\times \left(\int_0^1 \sum_{i=1}^{n-1} b_{ij} y_i^*(\tau)\,d\tau - \int_t^1 \sum_{i=1}^{n-1} b_{ij} y_i^*(\tau)\,d\tau \right) dt$$

$$= \int_0^1 x_j(t) \int_t^1 \sum_{i=1}^{n-1} b_{ij} y_i^*(\tau)\,d\tau\,dt. \tag{3.2.15}$$

If we substitute (3.2.14) and (3.2.15) in (3.2.13) we get

$$\langle Tx, y^* \rangle_{(n-1)} = \langle x, A^*y^* \rangle_{(n)} + \left\langle x, \int_t^1 B^*y^*(\tau)\,d\tau \right\rangle_{(n)}$$

$$= \left\langle x, A^*y^* + \int_t^1 B^*y^*(\tau)\,d\tau \right\rangle_{(n)}$$

thus

$$T^*y^* = A^*y^* + \int_t^1 B^*y^*(\tau)\,d\tau.$$

Let $a^* \in E^*$ and $b \in F$. Then the corresponding Lagrangian function takes the form

$$\psi(x, y^*) = -\left\langle y^*, Ax + \int_0^t Bx(\tau)\,d\tau \right\rangle_{(n-1)} + \langle a^*, b \rangle_{(n)} + \langle y^*, b \rangle_{(n-1)}$$

$$= -\left\langle x, A^*y^* + \int_t^1 B^*y^*(\tau)\,d\tau \right\rangle_{(n)} + \langle a^*, x \rangle_{(n)} + \langle y^*, b \rangle_{(n-1)}$$

defined on $P \times Q^* = \mathbf{L}_2{}^{n+} \times \mathbf{L}_2{}^{(n-1)+}$.

The construction of a dual pair of problems now follows:

$$\inf_{Q^*} \psi(x, y^*) = \langle a^*, x \rangle_{(n)} + \inf_{Q^*} \left\langle y^*, b - Ax - \int_0^t Bx(\tau)\,d\tau \right\rangle = \langle a^*, x \rangle_{(n)}$$

for all

$$x \in Z = \left\{x \in \mathbf{L}_2^{n+} : b - Ax - \int_0^t Bx(\tau)\,d\tau \in \mathbf{L}_2^{(n-1)+},\ 0 \leqq t \leqq 1\right\}.$$

$$\sup_P \psi(x, y^*) = \langle y^*, b\rangle_{(n-1)} + \sup_P \left\langle x, a^* - A^*y^* - \int_t^1 B^*y^*(\tau)\,d\tau\right\rangle_{(n)}$$

$$= \langle y^*, b\rangle_{(n-1)}$$

for all

$$y^* \in Z^* = \left\{y^* \in \mathbf{L}_2^{(n-1)+} : A^*y^* + \int_t^1 B^*y^*(\tau)\,d\tau - a^* \in \mathbf{L}_2^{n+},\ 0 \leqq t \leqq 1\right\}.$$

Thus the primary and dual problem take the form

$$W = \sup_Z \int_0^1 \sum_{i=1}^n a_i^*(t)\, x_i(t)\,dt, \tag{P}$$

$$W^* = \inf_{Z^*} \int_0^1 \sum_{i=1}^{n-1} y_i^*(t)\, b_i(t)\,dt \tag{D}$$

Obviously the convex cones P and Q are closed. Moreover it follows the sets of admissible points are convex and closed. We prove this assertion for Z. The convexity of Z is obvious. Now we show that the operator T is bounded, in the sense of $\|Tx\|_{(n-1)} \leqq c\,\|x\|_{(n)}$ for all $x \in \mathbf{L}_2^n[0, 1]$ whereby the constant c does not depend on x. As a matter of fact by application of the Minkowski inequality and $\bar{c} = \max_{i,j} \{a_{ji}, b_{ji}\}$ we get

$$\begin{aligned}\|Tx\|_{(n-1)} &= (\langle Tx, Tx\rangle_{(n-1)})^{1/2}\\ &= \left(\int_0^1 \sum_{j=1}^{n-1} \left(\sum_{i=1}^n a_{ji}x_i(t) + \sum_{i=1}^n b_{ji} \int_0^t x_i(\tau)\,d\tau\right)^2 dt\right)^{1/2}\\ &\leqq \sqrt{n-1}\,\bar{c} \left(\left(\int_0^1 \left(\sum_{i=1}^n x_i(t)\right)^2 dt\right)^{1/2} + \left(\int_0^1 \left(\sum_{i=1}^n \int_0^t x_i(\tau)\,d\tau\right)^2 dt\right)^{1/2}\right)\\ &\leqq c \left(\int_0^1 \sum_{i=1}^n x_i^2(t)\,dt\right)^{1/2} = c\,\|x\|_{(n)}.\end{aligned}$$

Now we consider any convergent sequence $\{x^k\} \subseteqq Z$. Since $x^k \in P$ and P is closed there exists an element $x \in P$ such that $\lim \|x^k - x\|_{(n)} = 0$. It is $b - Tx^k \in Q$. Since T is linear and bounded we get

$$\|(b - Tx^k) - (b - Tx)\|_{(n-1)} = \|Tx - Tx^k\|_{(n-1)} \leqq c\,\|x - x^k\|_{(n)},$$

thus $\lim_k (b - Tx^k) = b - Tx$. Since Q is a closed convex cone it follows $b - Tx \in Q$, hence Z is closed.

If we put in Example 4 for $n = 4$ and choose special constant elements a_{ij}, b_{ij} of the matrices A and B we get a model of conglomeration of steel and automobile production investigated by Bellman [1].

Before we present an other example we define:

Definition 3.2.3. The program $[T, P, Q, a^*, b]$ is said to be *self-dual* if $P = Q^*$, $Q = P^*$, $F = E^*$, $T = -T^*$, $b = -a^*$.

From the definition obviously follows according to the transformation given in Definition (3.2.3) for self-dual programs:

$$[T, P, Q, a^*, b]^{\circledast} = [-T^*, Q^*, P^*, -b, -a^*] = [T, P, Q, a^*, b].$$

Lemma 3.2.4. *If a self-dual program $[T, P, Q, a^*, b]$ is consistent the value W is finite. There exists an admissible sequence $\{x_k\}$ such that $\langle a^*, x_k\rangle = 0$.*

Proof: Since $P = Q^*$ from $b - Tx \in Q$ it follows $\langle b, x\rangle - \langle Tx, x\rangle \geqq 0$. According to $T = -T^*$ we get $\langle Tx, x\rangle = \langle x, T^*x\rangle = -\langle Tx, x\rangle$, consequently $\langle Tx, x\rangle = 0$. Because $b = -a^*$ it now follows $\langle a^*, x\rangle \leqq 0$ and this proves that W must be finite.

According to Theorem 3.2.5 we get $\omega^* + W$. By Theorem 3.2.3 there exists an admissible sequence $\{\bar{y}_k^*\} \in \hat{Z}^*$ such that $\lim_{n\to\infty} \langle y_n^*, b\rangle = \omega^* = W \leqq 0$. Because the program $[T, P, Q, a^*, b]$ is self-dual and $\{\bar{y}_k^*\} \in \hat{Z}^*$, i.e. $\lim_{n\to\infty} (T^*y_k^* - x_k^*) = a^*$, it follows

$$\lim_{n\to\infty} (Ty_k^* + x_k^*) = b \quad \text{thus } \bar{y}_k^* \in \hat{Z}.$$

In case $W = 0$ there is nothing more to prove. Now let $W < 0$. There exists an integer n_0 such that for all $n > n_0$ we get $\langle \bar{y}_k^*, b\rangle \leqq \omega^* = W < 0$ thus $\langle a^*, \bar{y}_k^*\rangle > 0$. Furthermore there exists an element $\bar{x} \in Z$ such that $\langle a^*, \bar{x}\rangle \leqq W < 0$. We consider the sequence $x_k = \delta_k\bar{y}_k^* + (1 - \delta_k)\,\bar{x}$ where δ_k is defined by $\delta_k\langle a^*, \bar{y}_k^*\rangle + (1 - \delta_k)\,\langle a^*, \bar{x}\rangle = 0$. The definition of δ_k implies

$$\delta_k = \frac{\langle a^*, \bar{x}\rangle}{\langle a^*, \bar{x}\rangle - \langle a^*, \bar{y}_k^*\rangle}$$

thus $0 < \delta_k < 1$. It is easy to be seen that the sequence $x_k = \delta_k\bar{y}_k^* + (1 - \delta_k)\,\bar{x}$ is admissible thus $\{x_k\} \in \hat{Z}$. Moreover we get for $n > n_0$

$$\langle a^*, x_k\rangle = \delta_k\langle a^*, \bar{y}_k^*\rangle + (1 - \delta_k)\,\langle a^*, \bar{x}\rangle$$

$$= \frac{\langle a^*, \bar{x}\rangle}{\langle a^*, \bar{x}\rangle - \langle a^*, \bar{y}_k^*\rangle}\,\langle a^*, \bar{y}_k^*\rangle + \left(\frac{-\langle a^*, \bar{y}_k^*\rangle}{\langle a^*, \bar{x}\rangle - \langle a^*, \bar{y}_k^*\rangle}\right)\langle a^*, \bar{x}\rangle = 0$$

and this completes the proof of the lemma. □

Lemma 3.2.5. *If a self-dual program $[T, P, Q, a^*, b]$ is superconsistent then this program is realizable and the value is zero.*

Proof: Regarding Lemma 3.2.4 it follows that the value of the program is finite. Then the proof of the lemma follows from Theorem 3.2.11. □

We will now present examples of self-dual programs.

Example 5. We consider

$$E = \{x(t)\colon x(t) \text{ denotes functions of bounded variation and}$$

$$x(0) = 0,\ 2x(t) = x(t + 0) + x(t - 0)\}.$$

Furthermore we choose $F = \mathbf{C}[0, 1]$ whereby $\mathbf{C}[0, 1]$ denotes the continuous functions defined on the closed interval $[0, 1]$. According to the theorem of F. Riesz we can present the linear continuous functionals defined on $F = \mathbf{C}[0, 1]$ in form of

$$\langle x, y\rangle = \int_0^1 y(t)\, \mathrm{d}x(t), \quad y \in F, \quad x \in E,$$

whereby the right-hand side denotes the Stieltjes integral. Thus E is isomorphic to F^* and $E^* = F$. The convex cone Q in F we define as

$$Q = \mathbf{C}^+ = \{y \in F: y(t) \geqq 0, 0 \leqq t \leqq 1\} = P^*.$$

Let $x \in E$ be a nondecreasing function. From definition of the Stieltjes integral it follows for all $y \in Q$, $\langle x, y\rangle \geqq 0$ thus the nondecreasing functions are contained in Q^*. On the other hand we assume that $x \in E$ has the property

$$\langle x, y\rangle = \int_0^1 y(t)\, \mathrm{d}x(t) \geqq 0 \quad \text{for all} \quad y \in Q.$$

Now we consider trapezoidal functions $y \in Q$ of unite altitude with $y(a) = y(b) = 1/2$, $a < b$ and $a, b \in [0, 1]$. Then as the sides of the trapezoides become steeper it follows

$$\lim \int_0^1 y(t)\, \mathrm{d}x(t) = \int_a^b \mathrm{d}x(t) = x(b) - x(a) \geqq 0.$$

Hence $x(t)$ is a nondecreasing function. Consequently the convex cone $Q^* = P \subseteqq E$ consists of the nondecreasing functions. Now let $\varphi(s, t)$ be a continuous function with the property $\varphi(s, t) = -\varphi(t, s)$. We define $T: E \to F$ in form of

$$Tx = \int_0^1 \varphi(t, s)\, \mathrm{d}x(s).$$

We calculate the dual operator $T^*: E \to F$ according to the relation $\langle Tx, y\rangle = \langle x, T^*y\rangle$ for all $(x, y) \in E \times F$. It is $\langle Tx, y\rangle = \left\langle \int_0^1 \varphi(t, s)\, \mathrm{d}x(s), y\right\rangle = \int_0^1 \left(\int_0^1 \varphi(t, s)\, \mathrm{d}x(s)\right) \mathrm{d}y(t)$ $= \int_0^1 \left(\int_0^1 -\varphi(s, t)\, \mathrm{d}y(t)\right) \mathrm{d}x(s) = \langle -Ty, x\rangle = \langle T^*y, x\rangle$ thus $T^* = -T$. If we choose $-a^* = b \in F$ then the constructed example $[T, P, Q, a^*, b]$ is a self-dual program. If there exists an element $x_0 \in P$ and $b \in F$ such that

$$b - Tx_0 = b - \int_0^1 \varphi(t, s)\, \mathrm{d}x(s) \in \mathbf{C}^+$$

is an interior point of $\mathbf{C}^+$ then the self-dual program $[T, P, Q, a^*, b]$ is superconsistent. According to Lemma 3.2.5 there exist an element $\overline{x} \in Z$, i.e. $\overline{x} \in P$ and $b - T\overline{x} \in \mathbf{C}^+$ with the property

$$W = \langle a^*, \overline{x}\rangle = -\int_0^1 \overline{x}(t)\, \mathrm{d}b(t) = 0.$$

3.3. Quadratic programming

Let T be a matrix of order (m, n) and $a \in \mathbb{R}^n$, $b \in \mathbb{R}^m$ be given vectors. Furthermore, let us consider a nonsingular matrix B of order (m, m), which is positive semidefinite and symmetric. Thus we have $B = B^*$ and for any $y \in \mathbb{R}^m$,

$$\langle y, By\rangle \geqq 0.$$

Because B is not singular, the equation $By = z$ is solvable for any $z \in \mathbb{R}^m$ and the solution is uniquely defined by $y = B^{-1}z$. Consequently, $\langle B^{-1}z, z\rangle \geqq 0$ for any $z \in \mathbb{R}^m$ so that B^{-1} is positive semidefinite also. The function $h(y) = \langle y, By\rangle$ is convex on $\mathbb{R}^m$. In fact, if we consider any real number λ, $0 \leqq \lambda \leqq 1$, and vectors $y^1, y^2 \in \mathbb{R}^m$, we have

$$\begin{aligned} h\big(\lambda y^1 + (1-\lambda)\, y^2\big) &= \big\langle \lambda y^1 + (1-\lambda)\, y^2, B\big(\lambda y^1 + (1-\lambda)\, y^2\big)\big\rangle \\ &= \langle y^2, By^2\rangle + 2\lambda\langle y^1 - y^2, By^2\rangle + \lambda^2\langle y^1 - y^2, B(y^1 - y^2)\rangle. \end{aligned}$$

Because B is positive semidefinite, we have

$$\lambda^2\langle y^1 - y^2, B(y^1 - y^2)\rangle \leqq \lambda\langle y^1 - y^2, B(y^1 - y^2)\rangle.$$

It immediately follows that

$$\begin{aligned} h\big(\lambda y^1 + (1-\lambda)\, y^2\big) &\leqq \langle y^2, By^2\rangle + 2\lambda\langle y^1 - y^2, By^2\rangle + \lambda\langle y^1 - y^2, B(y^1 - y^2)\rangle \\ &= \langle y^2, By^2\rangle + \lambda\langle y^1 - y^2, By^2\rangle + \lambda\langle y^1 - y^2, By^1\rangle \\ &= \lambda\langle y^1, By^1\rangle - \lambda\langle y^2, By^2\rangle + \langle y^2, By^2\rangle \\ &= \lambda h(y^1) + (1-\lambda)\, h(y^2). \end{aligned}$$

Now, we put

$$\begin{aligned} \varphi(x, y) &= -\langle Tx, y\rangle \\ \alpha(x) &= -\langle a, x\rangle \\ \beta(y) &= -\left(\langle b, y\rangle + \frac{1}{2}\langle y, By\rangle\right) \end{aligned}$$

and consider the Lagrangian form

$$\psi(x, y) = -\langle Tx, y\rangle + \langle a, x\rangle + \langle b, y\rangle + \frac{1}{2}\langle y, By\rangle$$

defined on $K \times L = (\mathbb{R}^{r+} \times \mathbb{R}^s) \times \mathbb{R}^m$.
First we calculate

$$\beta^*(x) = \inf_{\mathbb{R}^m}\left\{-\langle Tx, y\rangle + \langle b, y\rangle + \frac{1}{2}\langle y, By\rangle\right\} = \inf_{\mathbb{R}^m}\left\{\left\langle y, b - Tx + \frac{1}{2}By\right\rangle\right\}.$$

Because $\langle y, By\rangle$ is convex so also is the function $g(y) = \left\langle y, b - Tx + \frac{1}{2}By\right\rangle$ convex on $\mathbb{R}^m$ for all $x \in K = \mathbb{R}^{r+} \times \mathbb{R}^s$. Furthermore, $g(y)$ is obviously continuously differentiable and we get $(dg)_y = \operatorname{grad} g(y) = b - Tx + By$. According to Theorem 1.4.4, a relative minimum is a global minimum. If $y^0 \in \mathbb{R}^m$ is a vector with the property $(dg)_{y^0} = 0$, it follows from Theorem 1.4.10 immediately that g attains the minimum at y^0. In fact, according to Theorem 1.4.10, we have, for any $y \in \mathbb{R}^m$

$$g(y) - g(y^0) \geqq \langle (dg)_{y^0}, y - y^0\rangle = 0.$$

Consequently, the function g realized a global minimum for all those points $y \in \mathbb{R}^m$, for which

$$b - Tx + By = 0.$$

Because B is not singular, we have

$$y = B^{-1}(Tx - b)$$

and thus y is uniquely defined by x. With this value of y, we get

$$\beta^*(x) = g(B^{-1}(Tx - b)) = -\left\langle B^{-1}(Tx - b),\, Tx - b - \frac{1}{2}(Tx - b)\right\rangle$$

$$= -\frac{1}{2}\langle B^{-1}(Tx - b),\, Tx - b\rangle$$

where $\beta^*(x)$ is defined for all $x \in M = \mathbb{R}^{r+} \times \mathbb{R}^s$. To calculate $\alpha^*(y)$ with respect to $K = \mathbb{R}^{r+} \times \mathbb{R}^s$, we make the following decomposition:

$$a = (a^1, a^2) \in \mathbb{R}^r \times \mathbb{R}^s;\ x = (u, v) \in \mathbb{R}^{r+} \times \mathbb{R}^s;\ T = (T_1, T_2).$$

Then, we have

$$\langle a, x\rangle = \langle a^1, u\rangle + \langle a^2, v\rangle,$$

$$Tx = T_1 u + T_2 v,$$

$$T^*y = (T_1^*y, T_2^*y) \in \mathbb{R}^r \times \mathbb{R}^s.$$

Using these values,

$$\alpha^*(y) = \sup_{\mathbb{R}^{r+}\times\mathbb{R}^s} \{-\langle Tx, y\rangle + \langle a, x\rangle\}$$

$$= \sup_{\mathbb{R}^{r+}\times\mathbb{R}^s} \{\langle a^1, u\rangle + \langle a^2, v\rangle - \langle (u, v), (T_1^*y, T_2^*y)\rangle\}$$

$$= \sup_{\mathbb{R}^{r+}\times\mathbb{R}^s} \{\langle u, a^1 - T_1^*y\rangle + \langle v, a^2 - T_2^*y\rangle\}.$$

The supremum exists for all $y \in N$, where

$$N = \{y \in \mathbb{R}^m : T_1^*y - a^1 \in \mathbb{R}^{r+};\ a^2 - T_2^*y = 0\}.$$

For $y \in N$, we get $\alpha^*(y) = 0$. Thus, we finally have

$$\mu(x) = \inf_{\mathbb{R}^m} \varphi(x, y) = \beta^*(x) - \alpha(x) = \langle a, x\rangle - \frac{1}{2}\langle B^{-1}(Tx - b),\, Tx - b\rangle,$$

defined on $M = \mathbb{R}^{r+} \times \mathbb{R}^s$ and

$$\nu(y) = \sup_{\mathbb{R}^{r+}\times\mathbb{R}^s} \varphi(x, y) = \langle b, y\rangle + \frac{1}{2}\langle y, By\rangle,$$

defined on N.

Now the dual pair of problems appears in the following form:

$$\text{find } W = \sup_{\mathbb{R}^{r+}\times\mathbb{R}^s} \left\{\langle a, x\rangle - \frac{1}{2}\langle B^{-1}(Tx - b),\, Tx - b\rangle\right\} \qquad \text{(P)}$$

$$\text{find } W^* = \inf_N \left\{\langle b, y\rangle + \frac{1}{2}\langle y, By\rangle\right\}. \qquad \text{(D)}$$

The functions to be optimized in (P) and (D) involve quadratic forms which justify the name *Quadratic Programming*. Because B together with B^{-1} is positive semidefinite, the function $\mu(x)$ is concave while $\nu(y)$ is convex. From this point of view, quadratic programming is a special case of convex programming. In the context of quadratic programming Theorem 2.3.1 may now be rephrased in the form of the following theorem.

Theorem 3.3.1. *The following three statements are equivalent:*

(i) *There exist points* $(x^0, y^0) \in (\mathbb{R}^{r+} \times \mathbb{R}^s) \times \mathbb{R}^m$ *such that for any* $(x, y) \in (\mathbb{R}^{r+} \times \mathbb{R}^s) \times \mathbb{R}^m$, *we have*

$$\langle a, x\rangle + \left\langle y^0, b - Tx + \frac{1}{2} By^0\right\rangle \leqq \langle a, x^0\rangle + \left\langle y^0, b - Tx^0 + \frac{1}{2} By^0\right\rangle$$
$$\leqq \langle a, x^0\rangle + \left\langle y, b - Tx^0 + \frac{1}{2} By\right\rangle .$$

(ii) $N \neq \emptyset$ *and the problems* (P) *and* (D) *are realizable. Furthermore,*

$$W = \max_{\mathbb{R}^{r+}\times\mathbb{R}^s} \left\{\langle a, x\rangle - \frac{1}{2} \langle B^{-1}(Tx - b), Tx - b\rangle\right\}$$
$$= \min_N \left\{\langle b, y\rangle + \frac{1}{2} \langle y, By\rangle\right\} = W^*.$$

(iii) *There exist points* $(x^0, y^0) \in (\mathbb{R}^{r+} \times \mathbb{R}^s) \times \mathbb{R}^m$ *with*

$$\langle x^0, a - T^*y^0\rangle = 0,$$
$$\frac{1}{2} \langle B^{-1}(Tx^0 - b), Tx^0 - b\rangle = \left\langle y^0, Tx^0 - b - \frac{1}{2} By^0\right\rangle .$$

Now we represent the Lagrangian form φ in such a way that the terms involving only u and y separate out. Thus,

$$\varphi(x, y) = \varphi(u, v; y)$$
$$= -\langle T_1u + T_2v, y\rangle + \langle a^1, u\rangle + \langle a^2, v\rangle + \langle b, y\rangle + \frac{1}{2} \langle y, By\rangle$$
$$= \langle v, a^2 - T_2{}^*y\rangle + \chi(u, y)$$

where

$$\chi(u, y) = -\langle T_1u, y\rangle + \langle a^1, u\rangle + \left\langle y, b + \frac{1}{2} By\right\rangle .$$

If (x^0, y^0) are optimal programs for (P) and (D) and the statement (ii) is true, then $(x^0, y^0) = (u^0, v^0, y^0) \in (\mathbb{R}^{r+} \times \mathbb{R}^s) \times \mathbb{R}^m$ is a saddle point for φ; thus

$$\langle v, a^2 - T_2{}^*y^0\rangle + \chi(u, y^0) \leqq \langle v^0, a^2 - T_2{}^*y^0\rangle + \chi(u^0, y^0)$$
$$\leqq \langle v^0, a^2 - T_2{}^*y\rangle + \chi(u^0, y)$$

for any $(u, v, y) \in (\mathbb{R}^{r+} \times \mathbb{R}^s) \times \mathbb{R}^m$. Because $y^0 \in N$, we have $a^2 - T_2{}^*y^0 = 0$ and we get immediately

$$\chi(u, y^0) \leqq \chi(u^0, y^0) \leqq \chi(u^0, y) \tag{3.3.1}$$

for any $(u, y) \in K_1 \times L_1$ with $K_1 = \mathbb{R}^{r+}$ and

$$L_1 = \{y \in \mathbb{R}^m : a^2 - T_2{}^*y = 0\} .$$

Conversely, from the saddle-point inequality (3.3.1), it follows that N is a nonempty set. Then, regarding the inequality (2.3.3), it follows that

$$\langle a, x\rangle - \frac{1}{2}\langle B^{-1}(Tx - b), Tx - b\rangle \leqq \langle b, y\rangle + \frac{1}{2}\langle y, By\rangle$$

for any $(x, y) \in (\mathbb{R}^{r+} \times \mathbb{R}^s) \times N$. But it means that problems (P) and (D) are solvable. To show that the set N is not empty on the basis of the inequality (3.3.1), we consider a saddle point $(u^0, y^0) \in K_1 \times L_2$ for χ. Then $a^2 - T_2{}^*y^0 = 0$ holds. We have to show $T_1{}^*y^0 - a^1 \geqq 0$ which implies $y^0 \in N$. From (3.3.1) follows, for any $u \in \mathbb{R}^{r+}$

$$-\langle T_1 u, y^0\rangle + \langle a^1, u\rangle + \left\langle y^0, b + \frac{1}{2} By^0\right\rangle$$

$$\leqq -\langle T_1 u^0, y^0\rangle + \langle a^1, u^0\rangle + \left\langle y^0, b + \frac{1}{2} By^0\right\rangle.$$

Simplifying

$$\langle u - u^0, T_1{}^*y^0 - a^1\rangle \geqq 0. \tag{3.3.2}$$

We assume one component of vector $T_1{}^*y^0 - a^1$ is negative; for example the kth component $[T_1{}^*y^0 - a^1]_k < 0$. Then we put for the kth component of u,

$$[u]_k = [u^0]_k + n.$$

In this case,

$$[u - u^0]_k\,[T_1{}^*y^0 - a^1]_k = n[T_1{}^*y^0 - a^1]_k$$

and consequently,

$$\lim_{n\to\infty} n[T_1{}^*y - a^1]_k = -\infty.$$

But this is in contradiction to (3.3.2) which is true for any $u \in \mathbb{R}^{r+}$. Thus

$$T_1{}^*y^0 - a^1 \geqq 0.$$

If we respect the Remark 2 at the end of Section 2.4, we can apply Kuhn-Tucker condition to define a saddle point.

Now we put

$$\psi(x, y) = \psi(u, v; y) = -\langle T_1 u + T_2 v, y\rangle + \langle a^1, u\rangle + \langle a^2, v\rangle + \langle b, y\rangle + \frac{1}{2}\langle y, By\rangle$$

for which the condition (2.4.5),

$$\operatorname{grad}_y \psi(u^0, v^0; y^0) = 0, \quad \operatorname{grad}_u \psi(u^0, v^0; y^0) \leqq 0,$$
$$\operatorname{grad}_v \psi(u^0, v^0; y^0) = 0, \quad \langle \operatorname{grad}_u \psi(u^0, v^0; y^0), u^0\rangle = 0$$

takes the form.

$$b + By^0 - T_1 u^0 - T_2 v^0 = b + By^0 - Tx^0 = 0,$$
$$T_1{}^*y^0 - a^1 \in \mathbb{R}^{r+}, \ a^2 - T_2{}^*y^0 = 0$$
$$\langle u^0, a^1 - T_1{}^*y^0\rangle = 0.$$

This condition shows us that (u^0, v^0, y^0) is a saddle point if $y^0 \in N$ and

$$y^0 = B^{-1}(Tx^0 - b), \ \langle u^0, a^1 - T_1{}^*y^0\rangle = 0.$$

We can notice that demanding $y^0 \in N$ with $\langle u^0, a^1 - T_1{}^* y^0 \rangle = 0$ is equivalent to $\langle x^0, a - T^* y^0 \rangle = 0$ of statement (iii) in Theorem 3.3.1. In our investigation so far, it is assumed that B is a nonsingular matrix. But it is also possible to formulate a dual pair of problems in case B is singular. For this purpose, we consider the set $K = \mathbb{R}^{r+} \times \mathbb{R}^s$ and $L = \mathbb{R}^r \times \mathbb{R}^s$. Furthermore, let T be a matrix of order (r, s) and B be a positive semidefinite matrix of order (s, s) with $B^* = B$.

Using the given vectors $a \in \mathbb{R}^s$ and $b \in \mathbb{R}^r$, we put

$$\varphi(x, y) = \varphi(u, v; w, z) = -\langle w, u + Tv \rangle - \langle v, Bz \rangle,$$

$$\alpha(x) = \alpha(u, v) = -\langle a, v \rangle \quad \text{on } K = \mathbb{R}^{r+} \times \mathbb{R}^s,$$

$$\beta(y) = \beta(w, z) = -\langle b, w \rangle - \frac{1}{2} \langle z, Bz \rangle.$$

The Lagrangian form is

$$\psi(u, v; w, z) = -\langle w, u + Tv \rangle - \langle v, Bz \rangle + \langle a, v \rangle + \langle b, w \rangle + \frac{1}{2} \langle z, Bz \rangle$$

defined on $K \times L$.

We get the conjugate function β^* in the form,

$$\beta^*(x) = \inf_{\mathbb{R}^r \times \mathbb{R}^s} \left\{ -\langle w, u + Tv \rangle - \langle v, Bz \rangle + \langle b, w \rangle + \frac{1}{2} \langle z, Bz \rangle \right\}.$$

As we have done before, we can substitute infimum by minimum under the conditions

$$\big(\mathrm{d}(\varphi - \beta)\big)_w = -u - Tv + b = 0,$$

$$\big(\mathrm{d}(\varphi - \beta)\big)_z = -Bv + Bz = 0.$$

Then, we get for β^*,

$$\beta^*(x) = -\langle w, u \rangle - \langle v, T^* w + Bv \rangle + \langle b_2, w \rangle + \frac{1}{2} \langle v, Bv \rangle$$

$$= -\frac{1}{2} \langle v, Bv \rangle - \langle w, Tv + u - b \rangle = -\frac{1}{2} \langle v, Bv \rangle$$

defined on

$$M = \{(u, v) \in \mathbb{R}^{r+} \times \mathbb{R}^s \colon Tv + u = b\}.$$

For α^*, we get

$$\alpha^*(y) = \sup_{\mathbb{R}^{r+} \times \mathbb{R}^s} \{-\langle w, u + Tv \rangle - \langle v, Bz \rangle + \langle a, v \rangle\}$$

$$= \sup_{\mathbb{R}^{r+} \times \mathbb{R}^s} \{-\langle w, u \rangle - \langle v, T^* w + Bz - a \rangle\} = 0$$

with

$$y \in N = \{(w, z) \in \mathbb{R}^{r+} \times \mathbb{R}^s \colon T^* w + Bz = a\}.$$

As a pair of dual problems, we have

$$\text{find } W = \sup_M \left\{ \langle a, v \rangle - \frac{1}{2} \langle v, Bu \rangle \right\}, \tag{P}$$

$$\text{find } W^* = \inf_N \left\{ \langle b, w \rangle + \frac{1}{2} \langle z, Bz \rangle \right\}.$$

We notice that in (P), we have to optimize with respect to all v satisfying $Tv \leqq b$; because for a vector in M, the part u is not negative and $\beta^*(x) - \alpha(x)$ does not depend on u. Theorem 3.3.1 now appears in the following form.

Theorem 3.3.2. *The following three statements are equivalent:*

(i) *There exist points*

$$[(u^0, v^0), (w^0, z^0)] \in (\mathbb{R}^{r+} \times \mathbb{R}^s) \times (\mathbb{R}^r \times \mathbb{R}^s)$$

such that for any

$$[(u, v), (w, z)] \in (\mathbb{R}^{r+} \times \mathbb{R}^s) \times (\mathbb{R}^r \times \mathbb{R}^s)$$

we have

$$- \langle v, T^*w^0 + Bz^0 - a\rangle + \langle w^0, b - u\rangle + \frac{1}{2}\langle z^0, Bz^0\rangle$$

$$\leqq - \langle v^0, T^*w^0 + Bz^0 - a\rangle + \langle w^0, b - u^0\rangle + \frac{1}{2}\langle z^0, Bz^0\rangle$$

$$\leqq - \langle v^0, T^*w + Bz - a\rangle + \langle w, b - u^0\rangle + \frac{1}{2}\langle z, Bz\rangle.$$

(ii) *The sets M and N are not empty and the problems* (P) *and* (D) *are realizable. Furthermore,*

$$W = \max_M \left\{\langle a, v\rangle - \frac{1}{2}\langle v, Bv\rangle\right\} = \min_N \left\{\langle b, w\rangle + \frac{1}{2}\langle z, Bz\rangle\right\} = W^*.$$

(iii) *There exist points $[(u^0, v^0), (w^0, z^0)] \in M \times N$ such that*

$$\langle w^0, v^0\rangle = 0,$$

$$\langle w^0, u^0\rangle + \left\langle v^0, T^*w^0 + Bz^0 - \frac{1}{2} Bv^0\right\rangle = \langle b, w^0\rangle + \frac{1}{2}\langle z^0, Bz^0\rangle.$$

We leave it to the reader to verify that the Kuhn-Tucker conditions in this case have the form:

$$u^0 + Tv^0 = b, \quad Bv^0 = Bz^0, \quad T^*w^0 + Bz^0 = a,$$

$$w^0 \geqq 0, \quad \langle u^0, w^0\rangle = 0.$$

3.4. Convex programming

To construct the theory of convex programming, we refer to the studies made in Section 1.5 about conjugate convex functions. The theory of conjugate convex functions by Fenchel [12] gives us the possibility to construct a theory of convex programming in which the primal problem (P) can be formulated in point coordinates while the dual problem (D) appears in hyperplane coordinates according to the relations between a convex set $[f, C]$, associated with a convex function f and the corresponding dual set $[f, C]^*$. The results of inquiries made by Dieter [5] gives us the opportunity to represent different duality theorems of programming problems as special cases of a general duality principle, which Fenchel has already used for convex programming in 1953.

Let C and D be closed convex sets contained in $\mathbb{R}^n$, f a closed convex function defined on C and g a closed concave function defined on D. If g^* denotes the conjugate concave function corresponding to g defined on D^*, then we put

$$\varphi(x, y) = \langle a, x \rangle,$$
$$\alpha(x) = f(x),$$
$$\beta(a) = g^*(a),$$

and consider the Lagrangian form,

$$\psi(x, a) = \langle a, x \rangle - f(x) - g^*(a)$$

defined on $C \times D^*$. By f^* we mean the conjugate convex function corresponding to f with the domain of definition C^*. According to Lemma 1.5.1, the conjugate functions f^* and g^* are closed functions and from Lemma 1.5.2 follow $f^{**} = f$ and $g^{**} = g$. This is the case in Section 3.3 where we formulated Theorem 2.3.2.

The conjugate Lagrangian form ψ^* takes the form

$$\psi^*(x, a) = \langle a, x \rangle - f^*(a) - g(x)$$

defined on $D \times C^*$. Now we find

$$\mu(x) = \inf_{D^*} \psi(x, a) = g(x) - f(x), \quad x \in C \cap D,$$
$$\nu(a) = \sup_{C} \psi(x, a) = f^*(a) - g^*(a), \quad a \in C^* \cap D^*,$$
$$\mu^*(x) = \sup_{C^*} \psi^*(x, a) = f(x) - g(x), \quad x \in C \cap D,$$
$$\nu^*(a) = \inf_{D} \psi^*(x, a) = g^*(a) - f^*(a), \quad a \in C^* \cap D^*.$$

Corresponding to the Lagrangian form ψ and its conjugate ψ^*, we get two pairs of dual problems; namely

$$W_\psi = \sup_{C \cap D} \{g(x) - f(x)\}, \tag{P}$$
$$W_\psi^* = \inf_{C^* \cap D^*} \{f^*(a) - g^*(a)\}, \tag{D}$$
$$W_{\psi^*} = \sup_{C^* \cap D^*} \{g^*(a) - f^*(a)\}, \tag{P*}$$
$$W_{\psi^*}^* = \inf_{C \cap D} \{f(x) - g(x)\}. \tag{D*}$$

We can easily see that (P) is equivalent to (D*) and (D) is equivalent to (P*). If the sets $C \cap D$ and $C^* \cap D^*$ are not empty $g(x) - f(x) \leqq f^*(a) - g^*(a)$ holds for any $(x, a) \in C \cap D \times C^* \cap D^*$. From this result, it follows that all the four problems (P), (D), (P*), (D*) are solvable. In case $[f, C] \cap [g, D]$ is not empty, we can interpret W_ψ as diameter of this intersection in vertical direction while W_ψ is the distance between $[f, C]$ and $[g, D]$ in case $[f, C] \cap [g, D]$ is empty. To illustrate these ideas we may consider the following example:

Consider

$$f(x) = x^2, \quad \text{defined on } C = [-5, 5] \subseteqq \mathbb{R},$$
$$g(x) = -x^2 + 10, \quad \text{defined on } D = [-5, 5] \subseteqq \mathbb{R}.$$

Then

$$[f, C] = \{(\alpha, x) \in \mathbb{R} \times C : \alpha \geqq x^2\},$$
$$[g, D] = \{(\alpha, x) \in \mathbb{R} \times D : \alpha \leqq -x^2 + 10\}.$$

As conjugate functions, we have

$$f^*(a) = \sup_{[-5,5]} \{ax - x^2\} = \begin{cases} \dfrac{a^2}{4}, \ -10 \leqq a \leqq 10, \\ 5a - 25, \ 10 \leqq a, \\ -5a - 25, \ a \leqq -10, \end{cases}$$

$$g^*(a) = \inf_{[-5,5]} \{ax + x^2 - 10\} = \begin{cases} -\dfrac{a^2}{4} - 10, \ -10 \leqq a \leqq 10, \\ 5a + 10, \ a \leqq -10, \\ -5a + 15, \ 10 \leqq a \end{cases}$$

defined on $C^* = D^* = (-\infty, \infty) = \mathbb{R}$.

It is obvious that the problems (P) and (D) are realizable and

$$W_\psi = \max_{[-5,5]} \{-2x^2 + 10\} = \min_{\mathbb{R}} \{f^*(a) - g^*(a)\}$$
$$= \min_{[-10,10]} \left\{\frac{a^2}{2} + 10\right\} = W_\psi{}^*$$

i.e. $W = W^* = 10$ (see Fig. 6).

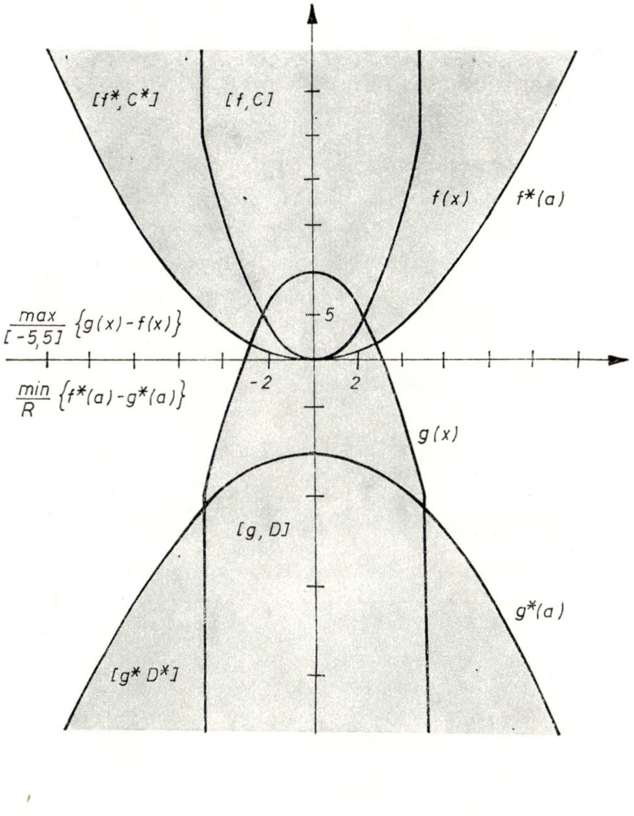

Fig. 6

In this special example, $[f, C] \cap [g, D] \neq \emptyset$ and $[f^*, C^*] \cap [g^*, D^*] = \emptyset$ As expected, $W_\psi = 10$ is a diameter in vertical direction of the nonempty intersection while $W_\psi{}^* = 10$ appears as a distance between $[f^*, C^*]$ and $[g^*, D^*]$. Of course, it depends on the special character of the example that the optimal programs of (P) and (D) are equal, namely $x = a = 0$. In the context of convex programming, Theorem 3.3.2 takes the following form:

Theorem 3.4.1. *The following statements are equivalent:*

(i) *There exist points* $(x^0, a^0) \in C \cap D \times C^* \cap D^*$ *such that*

$$\langle a^0, x\rangle - f(x) - g^*(a^0) \leqq \langle a^0, x^0\rangle - f(x^0) - g^*(a^0) \leqq \langle a, x^0\rangle - f(x^0) - g^*(a)$$

for any $(x, a) \in C \times D^*$ and

$$\langle a, x^0\rangle - f^*(a) - g(x^0) \leqq \langle a^0, x^0\rangle - f^*(a^0) - g(x^0) \leqq \langle a^0, x\rangle - f^*(a^0) - g(x)$$

for any $(x, a) \in D \times C^*$.

(ii) $C \cap D \neq \emptyset$ *and* $C^* \cap D^* \neq \emptyset$ *and*

$$W_\psi = \max_{C \cap D} \{g(x) - f(x)\} = \min_{C^* \cap D^*} \{f^*(a) - g^*(a)\} = W_\psi{}^*,$$

$$W_{\psi^*} = \max_{C^* \cap D^*} \{g^*(a) - f^*(a)\} = \min_{C \cap D} \{f(x) - g(x)\} = W_{\psi^*}^*$$

(iii) *There exist points* $(x^0, a^0) \in C \cap D \times C^* \cap D^*$ *such that*

$$\langle a^0, x^0\rangle = f(x^0) + f^*(a^0),$$

$$\langle a^0, x^0\rangle = g(x^0) + g^*(a^0).$$

According to (1.5.3), the statement (iii) means that the points (x^0, a^0) are conjugate points for f as well as for g. As we already know from Section 2.3, this statement is equivalent to $\mathcal{L}_f \cap \mathcal{L}_{g^*} \neq \emptyset$ where

$$\mathcal{L}_f = \{(x, a) \in C \times C^*: \langle a, x\rangle = f(x) + f^*(a)\},$$

$$\mathcal{L}_g{}^* = \{(x, a) \in D \times D^*: \langle a, x\rangle = g(x) + g^*(a)\}$$

The sets $\mathcal{L}_f$ and $\mathcal{L}_{g^*}$ just represent the community of all conjugate points for f and g respectively. Then it follows that the statements (i), (ii), (iii) of Theorem 3.4.1 are equivalent to the existence of points $(x^0, a^0) \in C \cap D \times C^* \cap D^*$ with the following properties:

(a) $\big(f^*(a^0), a^0\big) \in [f, C]^*$ produces a tangent hyperplane on $[f, C]$ through the point $\big(f(x^0), x^0\big) \in [f, C]$.

(b) $\big(f(x^0), x^0\big) \in [f, C]$ produces a tangent hyperplane on $[f, C]^*$ through the point $\big(f^*(a^0), a^0\big) \in [f, C]^*$.

(c) $\big(g^*(a^0), a^0\big) \in [g, D]^*$ produces a tangent hyperplane on $[g, D]$ through the point $\big(g(x^0), x^0\big) \in [g, D]$.

(d) $\big(g(x^0), x^0\big) \in [g, D]$ produces a tangent hyperplane on $[g, D]^*$ through the point $\big(g^*(a^0), a^0\big) \in [g, D]^*$.

In Theorem 3.4.1, we formulated an assertion about equivalence of certain different statements. For us, of course, existence conditions for at least one of the statements are also of interest. Therefore, we prove the following existence theorem.

Theorem 3.4.2. *If the sets $C \cap D$ and $C^* \cap D^*$ are not empty, then the problems* (P) *and* (D) *are solvable and we have*

$$\sup_{C \cap D} \{g(x) - f(x)\} = \inf_{C^* \cap D^*} \{f^*(a) - g^*(a)\}. \tag{3.4.1}$$

Proof: Solvability of problems (P) and (D) follows immediately from

$$g(x) - f(x) \leqq f^*(a) - g^*(a)$$

for any $(x, a) \in C \cap D \times C^* \cap D^*$.

To prove the relation (3.4.1) it is essential to apply Lemma 1.5.3. Consider $h(x) = f(x) + \big(-g(x)\big)$ which is a closed convex function defined on $C \cap D$. If $(-g^*)$ denotes the conjugate convex function corresponding to $-g$, we get

$$(-g)^* (a) = \sup_D \big\{\langle a, x\rangle - \big(-g(x)\big)\big\} = -\inf_D \{\langle -a, x\rangle - g(x)\} = -g^*(-a)$$

where g^* is the conjugate concave function corresponding to g. If D^* is the domain of definition for the conjugate g^*, then $-a \in D^*$, implies $a \in -D^*$ i.e. $-D^*$ is the domain of definition for $(-g^*)$. According to Lemma 1.5.3 the set $C^* + (-D^*) = C^* - D^*$ is contained in the domain of definition of the conjugate of function h. Because $C^* \cap D^* \neq \emptyset$, the zero element is contained in $C^* - D^*$, whence it follows that the conjugate h^* is defined for $a = 0$. On one hand, we get

$$h^*(0) = \sup_{C \cap D} \{\langle 0, x\rangle - h(x)\} = \sup_{C \cap D} \{g(x) - f(x)\}. \tag{3.4.2}$$

On the other hand, according to Lemma 1.5.3,

$$\begin{aligned} h^*(0) &= \inf \{f^*(a^1) + (-g^*)(a^2);\, a^1 + a^2 = 0,\ a^1 \in C^*, a^1 \in D^*\} \\ &= \inf_{C^* \cap D^*} \{f^*(a) - g^*(a)\} \end{aligned} \tag{3.4.3}$$

(3.4.2) and (3.4.3) together imply (3.4.1) and this proves the theorem. □

The statement (ii) of Theorem 3.4.1 differs from the relation (3.4.1) of Theorem 3.4.2. From (3.4.1) we only know that the values of the problems (P) and (D) are equal while from statement (ii) it follows that, except the equality of the values of problems (P) and (D), the problems are realizable and there exist optimal programs. In case of linear programming and quadratic programming, we can always substitute supremum by maximum and infimum by minimum, if the values of the programs are finite.

For convex programming, this is not true. Although the assumption of Theorem 3.4.2 that $C \cap D$ and $C^* \cap D^*$ are nonempty sets ensure that the values of the problems are finite it does not automatically imply realizability. It means that the assertion (3.4.1) cannot be improved. To explain about our remarks, we now give an example.

Let us consider sets $C = D = [1, \infty]$ and functions $f(x) = \frac{1}{x}$ defined on C, $g(x) = -\frac{1}{x}$ defined on D. Obviously f is convex and g concave. The conjugate functions f^* and g^* are now given by

$$f^*(a) = \begin{cases} -2\sqrt{-a}, & -1 \leqq a \leqq 0, \\ a - 1, & a \leqq -1, \end{cases}$$

defined on $C^* = (-\infty, 0]$ and

$$g^*(a) = \begin{cases} 2\sqrt{a}, & 0 \leqq a \leqq 1, \\ a+1, & 1 \leqq a, \end{cases}$$

defined on $D^* = [0, \infty)$. Obviously $C \cap D = [1, \infty) \neq \emptyset$ and $C^* \cap D^* = \{0\} \neq \emptyset$. Consequently, according to Theorem 3.4.2,

$$W = \sup_{C \cap D} \{g(x) - f(x)\} = \inf_{C^* \cap D^*} \{f^*(a) - g^*(a)\} = W^*.$$

In fact,

$$W = \sup_{[1,\infty)} \left\{\frac{-2}{x}\right\} = 0,$$

$$W^* = \inf_{\{0\}} \{-2\sqrt{-a} - 2\sqrt{a}\} = 0.$$

Obviously, in our example, the primal problem is not realizable where as the dual problem is realizable and $a^0 = 0$ is the optimal program. The graphical explanation of our example is given in Fig. 7.

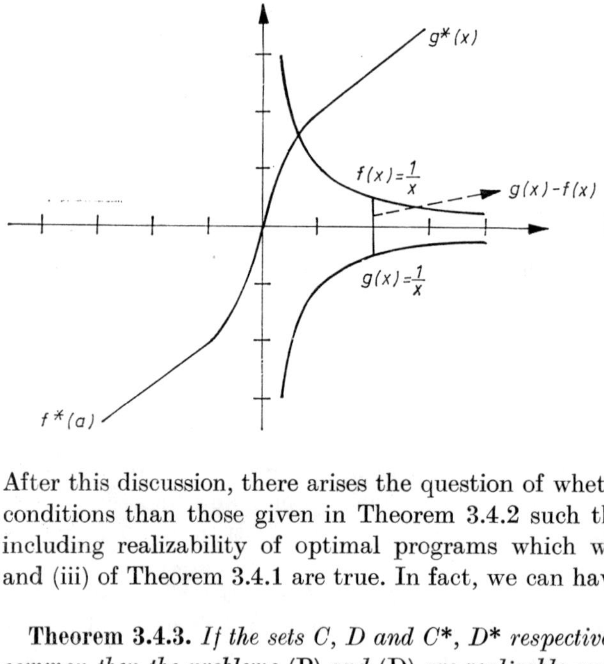

Fig. 7

After this discussion, there arises the question of whether it is possible to find stronger conditions than those given in Theorem 3.4.2 such that we get a theorem of duality including realizability of optimal programs which would ensure that statements (i) and (iii) of Theorem 3.4.1 are true. In fact, we can have such a theorem.

Theorem 3.4.3. *If the sets C, D and C^*, D^* respectively possess relative inner points in common then the problems* (P) *and* (D) *are realizable and*

$$\max_{C \cap D} \{g(x) - f(x)\} = \min_{C^* \cap D^*} \{f^*(a) - g^*(a)\}.$$

By a relative inner point in C, we mean an inner point in C with respect to the induced topology on C; see Edward [9].

Proof: Because $C \cap D \neq \emptyset$ and $C^* \cap D^* \neq \emptyset$, we have by Theorem 3.4.2,

$$\sup_{C \cap D} \{g(x) - f(x)\} = \inf_{C^* \cap D^*} \{f^*(a) - g^*(a)\}. \tag{3.4.4}$$

To prove the theorem we only have to show that the problems (P) and (D) are realizable. We can restrict our considerations to demonstrating the realizability of the dual problem (D). From the symmetry of Theorem 3.4.3 with respect to the functions f, g and f^*, g^* respectively follows the realizability of the primal problem (P). We put

$$W = \sup_{C \cap D} \{g(x) - f(x)\}$$

and define a new convex function

$$f_1(x) = f(x) + W$$

for which C is the domain of definition. Then the convex sets $[f_1, C]$ and $[g, D]$ have in common no inner points but at the most boundary points. Under this condition, we can apply Theorem 1.2.4, and consequently there exists a hyperplane H in $\mathbb{R}^{n+1}$ which separates these sets from each other. The hyperplane H must be a tangent hyperplane on $[f_1, C]$ as well as on $[g, D]$. Then there exist boundary points $\left(f_1(x^0), x^0\right) \in [f_1, C]$ and $\left(g(x^0), x^0\right) \in [g, D]$ with $\left(f_1(x^0), x^0\right) = \left(g(x^0), x^0\right)$ and the separating hyperplane H contains this boundary point. We know from Section 1.5 that the hyperplane H on $[f_1, C]$ through $\left(f_1(x^0), (x^0\right)$ can be produced by a boundary point $\left(f_1(a^0), a^0\right) \in [f_1, C^*]$ and on the other hand, H as a tangent hyperplane on $[g, D]$ through $\left(g(x^0), \eta^0\right)$ can be produced by $\left(g^*(a^0), a^0\right) \in [g, D]^*$. Obviously, the vector $(-1, a^0)$ defines the direction of H while $f_1(a^0) = g^*(a^0)$ is the cut made by H on the vertical axis. Thus,

$$\langle(-1, a^0), (\alpha, x)\rangle - f_1^*(a^0) \geqq 0$$

for any $(\alpha, x) \in [f_1, C]$

$$\langle(-1, a^0), (\alpha, x)\rangle - f_1^*(a^0) \leqq 0$$

for any $(\alpha, x) \in [g, D]$.

Furthermore,

$$\left\langle(-1, a^0), \left(f_1(x^0), x^0\right)\right\rangle - f_1^*(a^0) = -f_1(x^0) - f_1^*(a^0) + \langle a^0, x^0\rangle = 0.$$

From this relation follows

$$g^*(a^0) = f_1^*(a^0) = \sup_{C} \{\langle a^0, x\rangle - f(x) - W\}$$
$$= f^*(a^0) - W = \langle a^0, x^0\rangle - f(x^0) - W.$$

Consequently,

$$\sup_{C \cap D} \{g(x) - f(x)\} = W = f^*(a^0) - g^*(a^0) \geqq \inf_{C^* \cap D^*} \{f^*(a) - g^*(a)\}. \tag{3.4.5}$$

Because $a^0 \in C^* \cap D^*$, it follows from (3.4.4) and (3.4.5) that

$$f^*(a^0) - g^*(a^0) = \min_{C^* \cap D^*} \{f^*(a) - g^*(a)\}.$$

Thus the problem (D) is realizable and a^0 is an optimal program. To complete the proof, we must ensure that the separating hyperplane H is not vertical. Assume that H is vertical. Then, we consider the projections of the sets $[f_1, C]$ and $[g, D]$ as well as the projection of the hyperplane H on the space $\mathbb{R}^n$. It follows that C is the projection of $[f_1, C]$ and D is the projection of $[g, D]$ while the hyperplane H in the space $\mathbb{R}^{n+1}$ turns over to a hyperplane H' in $\mathbb{R}^n$. Because H is a separating hyperplane for the sets $[f_1, C]$ and $[g, D]$, the hyperplane H' must separate the sets C and D. This is in contradiction to our assumption that C and D possess relative inner points in common and thus the theorem is proved. □

Now we can draw the following conclusions:

If the sets C, D and C^*, D^* possess relative inner points in common then in view of Theorem 3.4.3, we have

$$\max_{C \cap D} \{g(x) - f(x)\} = \min_{C^* \cap D^*} \{f^*(a) - g^*(a)\}.$$

According to Theorem 3.4.1, this statement is equivalent to the existence of a saddle point $(x^0, a^0) \in C \cap D \times C^* \cap D^*$, satisfying

$$\psi(x, a^0) \leqq \psi(x^0, a^0) \leqq \psi(x^0, a)$$

for any $(x, a) \in C \times D^*$.

But the existence of a saddle point implies, according to Theorem 2.2.2, the relation,

$$\max_{C \cap D} \min_{D^*} \psi(x, a) = \min_{C^* \cap D^*} \max_{C} \psi(x, a).$$

In fact, Theorem 3.4.3 has a character of an existence theorem. If we change the assumption, we get a modification of this theorem in such a form that from solvability of the problem (P) follows solvability of problem (D). Theorems of this kind usually are called *duality theorems* in mathematical literature. Now, we prove a duality theorem.

Theorem 3.4.4. *If the sets C, D possess relative inner points in common and if there exists an optimal program for the primal problem* (P), *then there also exists an optimal program for the dual problem* (D), *and*

$$\max_{C \cap D} \{g(x) - f(x)\} = \min_{C^* \cap D^*} \{f^*(a) - g^*(a)\}.$$

Proof: Let $x^0 \in C \cap D$ be an optimal program for the primal problem (P). We put

$$W = g(x^0) - f(x^0) = \max_{C \cap D} \{g(x) - f(x)\}.$$

Similar to what we have done to prove Theorem 3.4.3 we separate the convex sets $[f_1, C]$ and $[g, D]$ by a suitable hyperplane where the direction of the normal to the hyperplane may be taken as the vector $(-1, a^0)$. The vector a^0 is contained in $C^* \cap D^*$ and we have

$$W = f^*(a^0) - g^*(a^0).$$

Furthermore, we already have the inequality

$$g(x) - f(x) \leqq f^*(a) - g^*(a)$$

for any $(x, a) \in C \cap D \times C^* \cap D^*$. □

Thus,

$$g(x^0) - f(x^0) = W = f^*(a^0) - g^*(a^0) \leqq f^*(a) - g^*(a)$$

for any $a \in C^* \cap D^*$.

Consequently, we have

$$\min_{C^* \cap D^*} \{f^*(a) - g^*(a)\} = f^*(a^0) - g^*(a^0) = W = \max_{C \cap D} \{g(x) - f(x)\}.$$

Lastly, we will demonstrate how we get linear programming and quadratic programming as special cases of convex programming. The approach is of interest especially for quadratic programming because, besides the equivalence statements of Theorem 3.2.1, we will also get conditions for existence of a solution from Theorems 3.4.3 and 3.4.4.

To construct linear programming as a special case of convex programming, we consider the space $U = \mathbb{R}^n \times \mathbb{R}^m$ and $V = \mathbb{R}^m \times \mathbb{R}^n$. Now we define a one to one linear mapping from U onto V of the form

$$S = \begin{bmatrix} T & I_m \\ -I_n & 0 \end{bmatrix}$$

where T is a matrix of order (m, n) and I_n, I_m are identity mappings from $\mathbb{R}^n$ onto $\mathbb{R}^n$ and $\mathbb{R}^m$ onto $\mathbb{R}^m$ respectively. 0 is a zero matrix. Now,

$$Su = \begin{bmatrix} T & I_m \\ -I_n & 0 \end{bmatrix} \begin{bmatrix} x \\ y \end{bmatrix} = \begin{bmatrix} Tx + y \\ -x \end{bmatrix} \in V$$

and

$$S^{-1}v = \begin{bmatrix} 0 & -I_n \\ I_m & T \end{bmatrix} \begin{bmatrix} y \\ x \end{bmatrix} = \begin{bmatrix} -x \\ Tx + y \end{bmatrix} \in U.$$

Let $a \in \mathbb{R}^n$ and $b \in \mathbb{R}^m$ be given vectors. Then $(a, 0) \in U$ and $(b, 0) \in V$. Furthermore, we consider convex cones in U and V of the form

$$P_u = \mathbb{R}^n \times \{0\} \subseteqq U, \quad Q_v = \mathbb{R}^{m+} \times \mathbb{R}^n \subseteqq V$$

and define $f(u) = 0$ with domain of definition

$$C = \{u \in U : (b, 0) - Su \in Q_v\}$$

as well as $g(u) = \langle (a, 0), u\rangle$ with domain of definition

$$D = \{u \in V : u \in P_u\}.$$

Obviously, C and D are convex sets and f is a convex function while g is concave. Now the primal problem of convex programming takes the form:

$$\text{Find } W = \max_{C \cap D} \{g(u) - f(u)\} = \max_{M} \{\langle a, x\rangle\} \tag{P}$$

with

$$C \cap D = \{u \in P_u : (b, 0) - Su \in Q_v\}$$

and

$$M = \{x \in \mathbb{R}^{n+} : b - Tx \in \mathbb{R}^{m+}\}.$$

But this problem is exactly problem (P_1) in Section 3.1.

To formulate the dual problem, we calculate the conjugate functions f^* and g^*. We notice that the transpose matrix S^* is a one to one linear mapping from V on U of the form

$$S^* = \begin{bmatrix} T^* & -I_n \\ I_m & 0 \end{bmatrix}.$$

Then, we have

$$\begin{aligned} f^*(u^*) &= \sup_C \{\langle u^*, u\rangle - f(u)\} = \sup_C \langle S^*v^*, u\rangle = \sup_C \langle v^*, Su\rangle \\ &= \sup_C \{\langle v^*, Su - (b, 0)\rangle + \langle v^*, (b, 0)\rangle\}. \end{aligned}$$

The supremum exists exactly for $u^* \in C^*$, where

$$C^* = \{u^* \in U : u^* = S^*v^*;\ v^* \in \mathbb{R}^{m+} \times \{0\}\}$$

and

$$f^*(u^*) = \langle v^*, (b, 0)\rangle .$$

In fact, if $v^* \in \mathbb{R}^{m+} \times \{0\}$, it follows that, for any $u \in C$,

$$\langle v^*, Su - (b, 0)\rangle \leqq 0 .$$

Because S is a one to one mapping the equation $Su - (b, 0) = 0$ possesses exactly one solution $u^0 = S^{-1}(b, 0)$ which is obviously contained in C. Thus $\sup \langle v^*, Su - (b, 0)\rangle = 0$. If, on the other hand, the element v^* is not contained in $\mathbb{R}^{m+} \times \{0\}$, then we choose an element $v^1 \in Q_v$ auch that $\langle v^*, v^1\rangle < 0$. Obviously, this is always possible. Now, we put $S^{-1}v^1 = u^1$ or $v^1 = Su^1$. Then follows

$$\langle v^*, Su^1\rangle < 0 . \tag{3.4.6}$$

For any $\alpha > 0$, the element $u^0 - \alpha u^1$ is contained in C, because

$$(b, 0) - S(u^0 - \alpha u^1) = \alpha v^1 \in Q_v .$$

Thus, $u^0 - \alpha u^1$ can be used in the calculation of the supremum. Consequently, in view of (3.4.6), we get

$$\sup_C \langle v^*, Su - (b, 0)\rangle = \lim_{\alpha \to \infty} \langle v^*, S(u^0 - \alpha u^1) - (b, 0)\rangle = \lim_{\alpha \to \infty} \{-\alpha \langle v^*, Su^1\rangle\} .$$

Thus, it follows that the defined set C^* is the domain of definition for function f^*. The conjugate concave function g^* takes the form

$$g^*(u^*) = \inf_D \{\langle u^*, u\rangle - \langle (a, 0), u\rangle\} = \inf_D \langle u, u^* - (a, 0)\rangle = 0$$

defined on

$$D^* = \{u^* \in U : u^* - (a, 0) \in \mathbb{R}^{n+} \times \mathbb{R}^m\} .$$

In fact, for any $u \in D$ and $u^* \in D^*$, $\langle u, u^* - (a, 0)\rangle \geqq 0$ and just because $u = 0$ is contained in D, we get $g^*(u^*) = 0$. Furthermore, the set D^* contains all elements for which the infimum exists. That is, if $u^* \notin D^*$, then there exists an element $u^1 \in D$ such that $\langle u^1, u^* - (a, 0)\rangle < 0$. Together with u^1, also the element αu^1, with $\alpha > 0$, is contained in D. Consequently, we get

$$\inf_D \langle u, u^* - (a, 0)\rangle = \lim_{\alpha \to \infty} \langle \alpha u^1, u^* - (a, 0)\rangle = -\infty .$$

Thus D^* is the domain of definition for g^*.

The dual problem of this special case of convex programming now takes the form:

$$\text{Find} \min_{C^* \cap D^*} \{f^*(u^*) - g^*(u^*)\} = \min_{C^* \cap D^*} \langle v^*, (b, 0)\rangle = \min_N \langle b, y\rangle \tag{D}$$

with

$$N = \{y \in \mathbb{R}^{m+} : T^*y - a \in \mathbb{R}^{n+}\} .$$

Here,

$$C^* \cap D^* = \{u^* \in U : u^* = S^*v^*;\ v^* \in \mathbb{R}^{m+} \times \{0\};\ u^* - (a, 0) \in \mathbb{R}^{n+} \times \mathbb{R}^m\}$$
$$= \left\{u^* \in U : u^* = \begin{bmatrix} T^*y \\ y \end{bmatrix};\ y \in \mathbb{R}^{m+};\ \begin{bmatrix} T^*y - a \\ y \end{bmatrix} \in \mathbb{R}^{n+} \times \mathbb{R}^m\right\}.$$

But, this is exactly the dual problem (D_1) of linear programming in Section 3.1. Theorems 3.4.1 and 3.4.4 imply the well known results of linear programming.

Next, we will demonstrate how to derive quadratic programming as a special case of convex programming. Here, beside, the equivalence Theorems 3.3.1 and 3.3.2, we get existence theorems for quadratic programming by applying existence theorems of convex programming. We consider a nonsingular positive semidefinite symmetric matrix B of order (m, m). Furthermore, let T be a matrix of order (m, n), and $a \in \mathbb{R}^n$, $b \in \mathbb{R}^m$ be given vectors. As we have seen before, the matrix B^{-1} is positive semidefinite as well, so that

$$f(u) = \langle a, x\rangle + \frac{1}{2}\langle B^{-1}y, y\rangle$$

is a convex function defined on $u = (x, y) \in C = \mathbb{R}^n \times \mathbb{R}^m$. Furthermore, let us consider a concave function $g(u) = 2\langle a,x\rangle$ defined on

$$D = \{(x, y) \in (\mathbb{R}^{r+} \times \mathbb{R}^s) \times \mathbb{R}^m : y = Tx - b\}.$$

Now we get

$$f^*(u^*) = \sup_{\mathbb{R}^n \times \mathbb{R}^m} \left\{\langle u, u^*\rangle - \langle a, x\rangle - \frac{1}{2}\langle B^{-1}y, y\rangle\right\}$$

$$= \sup_{\mathbb{R}^n \times \mathbb{R}^m} \left\{\langle x, x^* - a\rangle + \langle y, y^*\rangle - \frac{1}{2}\langle B^{-1}y, y\rangle\right\}.$$

Because x is any vector contained in $\mathbb{R}^n$ the supremum exists only if $x^* = a$.

Thus,

$$f^*(u^*) = \sup_{\mathbb{R}^m} \left\{\langle y, y^*\rangle - \frac{1}{2}\langle B^{-1}y, y\rangle\right\}.$$

The function

$$h(y) = \langle y, y^*\rangle - \frac{1}{2}\langle B^{-1}y, y\rangle$$

for any $y^* \in \mathbb{R}^m$, being concave, a relative maximum is a global maximum for the function h. If $y^0 \in \mathbb{R}^m$ is a vector with the property $(\mathrm{d}h)_{y^0} = 0$, it follows from Theorem 4.10 that

$$0 = \langle (\mathrm{d}h)_{y^0}, y - y^0\rangle \geqq h(y) - h(y^0)$$

i.e.

$$h(y) \leqq h(y^0) \quad \text{for any } y \in \mathbb{R}^m.$$

Then, the supremum will be attained for all y which satisfy the equation $(\mathrm{d}y)_y = y^* - B^{-1}y = 0$. Because B is nonsingular, the equation is solvable for any $y^* \in \mathbb{R}^m$ and the solution is $y = By^*$. Now, we have

$$f^*(u^*) = \langle By^*, y^*\rangle = \frac{1}{2}\langle y^*, By^*\rangle = \frac{1}{2}\langle B^*y^*, y^*\rangle$$

defined on $C^* = \{a\} \times \mathbb{R}^m$. The conjugate function g^* is now given by

$$g^*(u^*) = \inf_D \{\langle u, u^*\rangle - 2\langle a, x\rangle\} = \inf_{\mathbb{R}^{r+} \times \mathbb{R}^s} \{\langle x, x^*\rangle + \langle y^*, Tx - b\rangle - 2\langle a, x\rangle\}$$

$$= -\langle b, y^*\rangle + \inf_{\mathbb{R}^{t+} \times \mathbb{R}^s} \langle x, T^*y^* + x^* - 2a\rangle.$$

According to the representation of the space $\mathbb{R}^n = \mathbb{R}^r \times \mathbb{R}^s$, we get

$$\begin{aligned}\langle x, T^*y^* + x^* - 2a\rangle &= \langle (x^1, x^2), (T_1{}^*y^*, T_2{}^*y^*) + (x^{*1}, x^{*2}) - 2(a^1, a^2)\rangle \\ &= \langle x^1, T_1{}^*y^* + x^{*1} - 2a^1\rangle + \langle x^2, T_2{}^*y^* + x^{*2} - 2a^2\rangle .\end{aligned}$$

We observe that the infimum exists exactly for all $u^* = (x^*, y^*)$ with the property

$$T_1{}^*y^* + x^{*1} - 2a^1 \in \mathbb{R}^{r+},\ T_2{}^*y^* + x^{*2} - 2a^2 = 0 .$$

From this follows $g^*(u^*) = -\langle b, y^*\rangle$ defined on

$$D^* = \{(x^*, y^*) : T_1{}^*y^* + x^{*1} - 2a^1 \in \mathbb{R}^{r+};\ T_2{}^*y^* + x^{*2} - 2a^2 = 0\} .$$

Because C and D are closed convex sets and the functions f and g are closed convex and closed concave respectively, we have, according to Lemma 1.5.2, $f^{**} = f$, $g^{**} = g$, $C^{**} = C$, $D^{**} = D$. The Lagrangian form which produces a pair of dual quadratic programming problems has the form,

$$\varphi(u, u^*) = \langle u, u^*\rangle - \langle a, x\rangle - \frac{1}{2}\langle B^{-1}y, y\rangle + \langle b, y^*\rangle$$

defined on $C \times D^*$. The two problems are the following:

$$\begin{aligned}\text{Find } W &= \sup_{C \cap D}\{g(u) - f(u)\} = \sup_{D}\left\{\langle a, x\rangle - \frac{1}{2}\langle B^{-1}y, y\rangle\right\} \\ &= \sup_{\mathbb{R}^{r+}\times\mathbb{R}^s}\left\{\langle a, x\rangle - \frac{1}{2}\langle B^{-1}(Tx - b), Tx - b\rangle\right\},\end{aligned} \qquad \text{(P)}$$

$$\text{find } W^* = \inf_{C^* \cap D^*}\{f^*(u^*) - g^*(u^*)\} = \inf_{N}\left\{\langle b, y^*\rangle + \frac{1}{2}\langle By^*, y^*\rangle\right\} \qquad \text{(D)}$$

with

$$\begin{aligned}C^* \cap D^* &= (\{a\} \times \mathbb{R}^m) \cap \{(x^*, y^*) : T_1{}^*y^* + x^{*1} - 2a^1 \in \mathbb{R}^{r+};\ T_2{}^*y^* + x^{*2} - 2a^2 = 0\} \\ &= \{(a, y^*) : T_1{}^*y^* - a^1 \in \mathbb{R}^{r+};\ T_2{}^*y^* - a^2 = 0\} = N .\end{aligned}$$

This is the pair of dual problems already given in Section 3.3. The set $C \cap D$ is not empty. If we assume $N \neq \emptyset$, then we have, for any $(x, y^*) \in (\mathbb{R}^{r+}, \mathbb{R}^s) \times N$,

$$\langle a, x\rangle - \frac{1}{2}\langle B^{-1}(Tx - b), Tx - b\rangle \leqq \langle b, y^*\rangle + \frac{1}{2}\langle By^*, y^*\rangle .$$

Thus the values W and W^* are finite and because the functions to be optimized are quadratic, we can replace supremum by maximum and infimum by minimum.

Now, by Theorem 3.4.2, we have the following theorem.

Theorem 3.4.5. *If $N \neq \emptyset$, then the problems* (P) *and* (D) *of quadratic programming are solvable and realizable and the statements* (i), (ii), (iii) *of Theorem* 3.3.1 *are true.*

On the other hand, we can apply Theorem 3.4.1 on the special case of quadratic programming. The second statements of Theorems 3.3.1 and 3.4.1 are identical. Thus all statements (i), (ii), (iii) of Theorems 3.3.1 and 3.4.1 are true in case N is a nonempty set.

In conclusion, let us consider the Lagrangian form

$$\varphi_1(x, y) = \langle a, x\rangle + \left\langle y, b - Tx + \frac{1}{2} By\right\rangle$$

defined on $K \times L = (\mathbb{R}^{r+} \times \mathbb{R}^s) \times \mathbb{R}^m$ which produces the dual pair of quadratic programming in Section 3.3. According to the decomposition $\mathbb{R}^{r+} \times \mathbb{R}^s$, we could represent φ_1 by

$$\varphi_1(x, y) = \varphi_1(x^1, x^2; y) = \langle x^2, a^2 - T_2{}^*y\rangle + \chi(x^2, y)$$

with

$$\chi(x^1, y) = \langle a^1, x^1\rangle + \left\langle y, b - T_1x^1 + \frac{1}{2} By\right\rangle.$$

In view of Section 3.3, there is the following relationship between φ_1 and χ:

(z_1) If N is a nonempty set and if the duality statement (ii) of Theorem 3.3.1 is true then there exists a saddle point of χ with respect to

$$K_1 \times L_1 = \mathbb{R}^{r+} \times \{y \in \mathbb{R}^m\colon T_2{}^*y - a^2 = 0\}.$$

(z_2) If there exists a saddle point of χ with respect to $K_1 \times L_1$ then N is a nonempty set.

Furthermore, if we consider the function

$$\varphi_2(u, u^*) = \langle u, u^*\rangle - \langle a, x\rangle - \frac{1}{2}\langle B^{-1}y, y\rangle + \langle b, y^*\rangle$$

defined on

$$C \times D^* = (\mathbb{R}^n \times \mathbb{R}^m) \times \{(x^*, y^*)\colon T_1{}^*y^* + x^{*2} - 2a^1 \in \mathbb{R}^{r+};\ T_2{}^*y + x^{*2} - 2a^2 = 0\},$$

we get the following theorem:

Theorem 3.4.6. *The following four statements are equivalent:*

(i) *There exists a saddle point* (u^0, u^{*0}) *of* φ_2 *with respect to* $C \times D^*$ *with* $(u^0, u^{*0}) \in C \cap D \times C^* \cap D^*$.

(ii) *There exists a saddle point of* φ_1 *with respect to* $K \times L$.

(iii) *There exists a saddle point of* χ *with respect to* $K_1 \times L_1$.

(iv) N *is a nonempty set.*

The proof of the theorem follows from Theorem 3.4.5 and the results given in Section 3.3.

3.5. Duality of the transportation and potential problem in finite directed graphs

Very often, the mathematical theory of the transportation and potential problem in finite directed graphs is applied to describe some concrete practical problems of finding optimal solutions which eventually leads to rational decisions. The model of a network usually is a set of points (called vertices) and some arcs connecting some of the points. We come across these models as a simplex in the mathematical subject of topology, as an electrical circuit in physics as a diagram of organization in management and economy, as a representation of telephone network, as a system of pipelines, as road maps and in many other spheres of our daily life.

In this section, first we will give a short survey of the main ideas of theory of graphs that we need later in the development of our theory. For details we refer to BERGE [3]. It is not important for us to demonstrate all the relations between these basic concepts. We only put our concentration on those relations which are relevant to the construction of transportation and potential problem from the point of view of duality theory in order to produce a Lagrangian form. We will therefore present only those theorems and relations of graph theory so far as it is necessary to construct the theory of duality for this special case.

In the following, some basic facts of graph theory are introduced in such a way that the reader can follow them without any difficulty even if one is not well versed in this field of mathematics.

Let us consider a finite set X where we call elements of X the vertices of a graph. Furthermore, let U be a finite set of ordered pairs (a, b) with $a \in X$, $b \in X$. We call elements $(a, b) \in U$ the arcs of a graph.

The pair of sets $G = (X, U)$ is defined as a *finite directed graph.*

Such a graph is finite because the sets X and U are finite and it is directed because every element $(a, b) \in U$ can be represented as a directed arc going from a to b. We denote by $|X| = m$ the number of vertices and by $|U| = n$, the number of arcs. A *subgraph* (A, B) of G is a graph with $A \subseteqq X$ and B containing all those arcs from U which connect vertices contained in A.

Now, using a one to one mapping, we assign to each arc $(a, b) \in U$ one of the numbers $i \in \{1, 2, \ldots, n\}$; thus $(a, b) \to i$. From this point of view, we may identify the number i with the arc (a, b).

A chain is a series of consecutive arcs.

We call $\mu = \{i_1, \ldots, i_s\} \subseteqq \{1, \ldots, n\} = U$ a *cycle* of G if

1. μ is a chain.
2. $i_k \neq i_r$ if $k \neq r$; $1 \leqq k, r \leqq s$.
3. The arcs i_s and i_1 are consecutive.

Every cycle is oriented according to the given order of the arcs $i_1, \ldots, i_s$. If you traverse through a cycle μ according to its orientation, we pass some arcs in the direction of their orientation and some in the opposite direction of their orientation. We denote by μ^+ the set of all arcs we pass in the direction of their orientation while μ^- denotes the set of those arcs we pass in the opposite direction of their orientation.

Then, we can represent a cycle by $\mu = \mu^+ \cup \mu^-$.

A cycle μ is called *elementary* if no vertex in the cycle is encountered more than once except for the vertex which is both the initial and the final one.

Let A be a subset of X. By $\omega^+(A)$, we denote the set of all arcs from U which have the initial vertex in A and the terminal vertex not in A. The set $\omega^-(A)$ contains all arcs which have the initial vertex not in A and the terminal vertex in A.

If at least one of the sets $\omega^+(A)$ or $\omega^-(A)$ is not empty, we call

$$\omega(A) = \omega^+(A) \cup \omega^-(A)$$

a *co-cycle* of $G = (X, U)$.

Given a vertex $x \in X$ and A_x the set of all vertices which can be connected to x by a chain to which x itself is added. We call a subgraph formed by a set A_x a *connected component* of G. If G possesses only one connected component, we call G *connected.*

A co-cycle ω is said to be elementary if ω is the set of arcs joining two connected subgraphs A_1 and A_2 which have the following properties:

$$A_1 \neq \emptyset, \quad A_2 \neq \emptyset,$$

$$A_1 \cap A_2 = \emptyset,$$

$$A_1 \cup A_2 = A \subset X$$

where A is a connected component of G.

Now, we consider vectors $x = (\xi_1, \xi_2, \ldots, \xi_n) \in \mathbb{R}^n$ associated with $G = (X, U)$ where we assign to each arc $i \in U$ the component ξ_i. Especially, we can assign to each cycle μ and co-cycle $\omega(A)$ vectors $x(\mu)$ and $x(\omega)$ respectively by the definitions

$$x(\mu) = \big(\xi_1(\mu), \ldots, \xi_n(\mu)\big),$$

$$x(\omega) = \big(\xi_1(\omega), \ldots, \xi_n(\omega)\big)$$

with

$$\xi_i(\mu) = \begin{cases} 1, & \text{if } i \in \mu^+, \\ -1, & \text{if } i \in \mu^-, \\ 0, & \text{if } i \notin \mu, \end{cases}$$

$$\xi_i(\omega) = \begin{cases} 1, & \text{if } i \in \omega^+(A), \\ -1, & \text{if } i \in \omega^-(A), \\ 0, & \text{if } i \notin \omega(A), \end{cases}$$

for $i = 1, 2, \ldots, n$.

Let us consider the cycles $\mu_1, \ldots, \mu_k$ of G. We call $\mu_1, \ldots, \mu_k$ a *cycle basis* if

(i) $\mu_1, \ldots, \mu_k$ are elementary,

(ii) $x(\mu_1), x(\mu_2), \ldots, x(\mu_i)$ are linearly independent,

(iii) for any cycle μ of G there exist real numbers $\lambda_1, \lambda_2, \ldots, \lambda_k$ such that

$$x(\mu) = \sum_{i=1}^{k} \lambda_i x(\mu_k).$$

Analogously, we define co-cycles $\omega_1, \ldots, \omega_l$ to be a *co-cycle basis* if

(i) $\omega_1, \ldots, \omega_l$ are elementary,

(ii) $x(\omega_1), \ldots, x(\omega_l)$ are linearly independent,

(iii) for any co-cycle $\omega(A)$ of G there exist real numbers $\sigma_1, \sigma_2, \ldots, \sigma_l$ such that

$$x(\omega) = \sum_{i=1}^{l} \sigma_i x(\omega_i).$$

There exist methods to construct a cycle basis and a co-cycle basis of a given graph, but we do not discuss those methods here; for such methods refer Berge and Ghouila-Houri [4]. We observe that cycle and co-cycle bases of a given finite directed graph are not uniquely defined, but since a finite graph possesses at the most a finite number of cycles and co-cycles, there can exist only a finite number of cycle bases and co-cycle bases. On the other hand, as we have expected, the numbers k and l, the length of cycle and co-cycle bases are constants of the given graph. If we denote by p the number of connected components of a given graph G, then we have the following theorem.

Theorem 3.5.1. *Let $\mu_1, \ldots, \mu_k$ be a cycle basis and $\omega_1, \ldots, \omega_l$ be a co-cycle basis. If*

$$\Phi = L\big(x(\mu_1), \ldots, x(\mu_k)\big),$$
$$\Theta = L\big(x(\omega_1), \ldots, x(\omega_l)\big)$$

are the linear hulls of the cycle basis and the co-cycle basis respectively, then the dimensions of these subspaces are

$$k(G) = \dim \Phi = n - m + p,$$
$$l(G) = \dim \Theta = m - p.$$

The numbers $k(G)$ and $l(G)$ respectively are called *cyclomatic* and *co-cyclomatic numbers.*

Proof: The theorem is proved in three steps:

(1) there exist $m - p$ linearly independent elementary co-cycles
(2) there exist $n - m + p$ linearly independent elementary cycles
(3) there do not exist more than $n - m + p$ linearly independent cycles and $m - p$ linearly independent co-cycles.

(1) First we assume $p = 1$ and we construct $m - 1$ linearly independent co-cycles as follows: Let a_1 be any vertex contained in X. If we put $A_1 = \{a_1\}$, then the co-cycle $\omega(A_1)$ contains an elementary co-cycle. Let (a_1, a_2) or (a_2, a_1) with $a_1 \in A_1$ and $a_2 \notin A_1$ be an arc of this elementary co-cycle. We put $A_2 = A_1 \cup \{a_2\}$ and the co-cycle $\omega(A_2)$ contains an elementary co-cycle. From this elementary co-cycle, we choose an arc (a, a_3) or (a_3, a) with $a \in A_2$ and $a_3 \notin A_2$. We put $A_3 = A_2 \cup \{a_3\}$ and continue the process. Because the graph $G = (X, U)$ possesses m vertices we get from this construction $m - 1$ linearly independent elementary co-cycles. The independence of the co-cycles follows because each co-cycle contains an arc which is not contained in the next following one. If $p > 1$ and $A_1, \ldots, A_p$ are the connected components of G, we can apply this construction for each component A_i. Altogether, we get

$$\sum_{i=1}^{p} (|A_i| - 1) = \sum_{i=1}^{p} |A_i| - p = m - p$$

linearly independent elementary co-cycles.

(2) Let us consider graphs

$$G_0 = (X, \emptyset), G_1 = (X, \{1\}), \ldots, G_i = (X, \{1, 2, \ldots, i\}) \ldots, G_n = (X, U) = G$$

and a function

$$h(G_i) = i - m + p(i).$$

Here $p(i)$ denotes the number of components of G_i. Then,

$$h(G_0) = 0 - m + m = 0,$$
$$h(G_1) = 1 - m + (m - 1) = 0,$$
$$\vdots \qquad \vdots$$
$$h(G_n) = n - m + p(n) = n - m.$$

If we add to G_{i-1} the arc with the number i then we get G_i and there two possible relations between $h(G_i)$ and $h(G_{i-1})$. Either the arc i produces a cycle μ_i in which case G_i possesses one more arc than G_{i-1} while the number of components does not change and thus

$$h(G_i) = h(G_{i-1}) + 1$$

or the arc i does not produce a cycle in which case G_i possesses one more arc than G_{i-1} but the number of its components is one less and thus

$$h(G_i) = h(G_{i-1}).$$

From this follows that at the end of this procedure we have $h(G_n) = h(G) - n - m + p$ cycles $\mu_{i_1}, \ldots, \mu_{i_{n-m+p}}$.

These cycles are linearly independent, because

$$\alpha_1 x(\mu_{i_1}) + \cdots + \alpha_t x(\mu_{i_t}) = 0$$

with $1 \leqq t \leqq n - m + p$ and $\alpha_t \neq 0$ is not possible since the arc i_t is contained in μ_{i_t}, but not in any one of the cycles $\mu_{i_1}, \ldots, \mu_{i_{t-1}}$. Consequently, there also exist $n - m + p$ linearly independent elementary cycles.

(3) Let μ be any cycle and $\omega(A)$ any co-cycle of G. It is observed that

$$\Big\langle x(\mu), x\big(\omega(A)\big)\Big\rangle = \Big\langle x(\mu), \sum_{a \in A} x\big(\omega(a)\big)\Big\rangle = \sum_{a \in A} \Big\langle x(\mu), x\big(\omega(a)\big)\Big\rangle = 0. \tag{3.5.1}$$

If M is a subspace of $\mathbb{R}^n$ produced as a linear hull of all cycles of G and N a subspace produced as a linear hull of all co-cycles of G, then it follows from (3.5.1) that M and N are orthogonal. Thus, $\dim M + \dim N \leqq n$.

On the other hand, it follows from steps (1) and (2) that $\dim M + \dim N \geqq (n - m + p) + (m - p) = n$. Consequently $\dim M + \dim N = n$. Thus

$$k(G) = \dim M = \dim \Phi = n - m + p,$$

$$l(G) = \dim N = \dim \Theta = m - p.$$

Furthermore, the spaces Φ and Θ are orthogonal complements of $\mathbb{R}^n$ and this proves Theorem 3.5.1.

Now, we call a vector $x \in \mathbb{R}^n$ a *flow* over $G = (X, U)$ if for any $a \in X$, we have

$$\Big\langle x, x\big(\omega)(a)\big)\Big\rangle = \sum_{i \in \omega^+} \xi_i \xi_i(\omega) - \sum_{i \in \omega^-} \xi_i \xi_i(\omega) = 0.$$

We call a vector $z \in \mathbb{R}^n$ a *tension* over G, if for any elementary cycle μ, we have

$$\langle z, x(\mu)\rangle = \sum_{i \in \mu^+} \zeta_i \zeta_i(\mu) - \sum_{i \in \omega^-} \zeta_i \xi_i(\mu) = 0.$$

With respect to these definitions it is obvious that the set of all flows over G and the set of all tensions over G are linear subspaces of $\mathbb{R}^n$. The reader can easily ensure that $x(\mu)$ for any cycle μ is a flow and $x\big(\omega(A)\big)$ for any co-cycle $\omega(A)$ is a tension. Thus all elements of Φ are flows and all elements of Θ are tensions over G. If, for example, $u \in \mathbb{R}^n$ would be a flow not contained in Φ then it follows from the fact that flows and tensions are orthogonal to each other that

$$\Big\langle u, x\big(\omega(a)\big)\Big\rangle = 0$$

for any $a \in X$. Thus u is orthogonal to the whole subspace Θ. But the only vector not contained in Φ and orthogonal to Θ is the zero vector. This means Φ is the community of all flows and Θ the community of all tensions over G.

Now we enumerate all vertices of G from (1) to m so that $X = \{a_1, \ldots, a_m\}$. The matrix

$$S = [s_{ji}] = \begin{bmatrix} x(\omega(a_1)) \\ \vdots \\ x(\omega(a_m)) \end{bmatrix}$$

is of order (m, n) and we call S the *incident matrix* of the graph G. The matrix S describes in which way the vertices and arcs of G are connected to each other. Every column of S possesses only two elements unequal to zero, namely $+1$ and -1. If there is an arc $i = (a_k, a_r)$ then $s_{ki} = 1$ and $s_{ri} = -1$ and the remaining elements in column number i are zero. From the definition of a flow, it follows immediately that $x \in \mathbb{R}^n$ is a flow if and only if $Sx = 0$. A relation between a tension and matrix S is given by the following lemma.

Lemma 3.5.1. *A vector $z \in \mathbb{R}^n$ is a tension if and only if there exists a vector $y \in \mathbb{R}^m$ such that*

$$z + S^*y = 0. \tag{3.5.2}$$

If $z = (\zeta_1, \ldots, \zeta_n)$ and $y = (\eta_1, \ldots, \eta_m)$ and furthermore $i = (a_k, a_r) \in U$ is an arc of G, then because $s_{ki} = 1$ and $s_{ri} = -1$, it follows from (3.5.2) that

$$\zeta_i = \eta_r - \eta_k .$$

If we assign to each vertex a_k the component η_k of y, Lemma 3.5.1 means that z is a tension if and only if every component of z is the difference of the values associated with its vertices. From this point of view, we can interpret y as a potential corresponding to the tension z and the components of z are potential differences.

Proof of Lemma 3.5.1: First, we will show that z is a tension under the assumption that there exists a vector $y \in \mathbb{R}^m$ such that $z + S^*y = 0$. Let us consider any elementary cycle μ. It follows from (3.5.2) that

$$\langle z, x(\mu)\rangle = -\langle S^*y, x(\mu)\rangle = -\langle y, Sx(\mu)\rangle .$$

Since the vector $x(\mu)$ is a flow for any cycle μ, we have $Sx(\mu) = 0$ and thus

$$\langle z, x(\mu)\rangle = 0 .$$

Consequently z is a tension.

Next we assume that z is a tension. To prove the lemma, we must show the existence of a corresponding potential $y \in \mathbb{R}^m$. We may construct y as follows.

Let a and b be two vertices of a connected component of G. Then there exists at least one chain from a to b. We denote this chain by $\Pi_1 = (i_1, \ldots, i_s)$ and consider the vector $x(\Pi_1) \in \mathbb{R}^n$ which is defined analogous to corresponding vectors for cycles.

If we assign to the vertex a the value $y(a) = 0$, then we define the corresponding value for the vertex b by

$$y(b) = \langle z, x(\Pi_1)\rangle .$$

z being a tension the value $y(b)$ is uniquely defined. Because, if $\Pi_2 = (j_1, \ldots, j_r)$ is another chain from a to b and

$$\langle z, x(\Pi_2)\rangle \neq y(b)$$

then, in this case, we get

$$\langle z, x(\Pi_1) - x(\Pi_2)\rangle \neq 0 .$$

But the vector $x(\Pi_1) - x(\Pi_2)$ is obviously the corresponding vector for the cycle $\{i_1, \ldots, i_s, j_r, \ldots, j_1\}$ and therefore $\langle z, x(\Pi_1) - x(\Pi_2)\rangle \neq 0$ contradicts the assumption that z is a tension.

Consequently, the construction produces a uniquely defined vector $y = (y(a_1), \ldots, y(a_m))$ which is a potential for z and thus y satisfied (3.5.2).

Here ends our preparations on graph theory, which we need to construct a duality theory for transportation and potential problems.

Let $G(X, U)$ be a given finite directed graph. We assign to each arc $i \in U$ a closed interval $J_i = [b_i, c_i]$, $i = 1, 2, \ldots, n$. Furthermore, let f_i be convex functions defined on J_i. Obviously, the function

$$F(x) = \sum_{i=1}^{n} f_i(\xi_i)$$

is a convex function defined on $Q = J_1 \times \cdots \times J_n$. If f_i^* are the corresponding conjugate convex functions defined on $J_i^* = [b_i^*, c_i^*]$, then we get

$$\begin{aligned}\sup_Q \{\langle x, z\rangle - F(x)\} &= \sup_Q \left\{\sum_{i=1}^{n} \left(\xi_i\zeta_i - f_i(\xi_i)\right)\right\} \\ &= \sum_{i=1}^{n} \sup_{J_i} \{\xi_i\zeta_i - f_i(\xi_i)\} = \sum_{i=1}^{n} f_i^*(\zeta_i) = F^*(z)\end{aligned}$$

where F^* is the conjugate convex function corresponding to F defined on $Q^* = J_1^* \times \cdots \times J_n^*$.

Now, we put

$$\begin{aligned}\varphi(x, y) &= -\langle Sx, y\rangle \\ \alpha(x) &= F(x) \\ \beta(y) &= 0\end{aligned}$$

and construct a Lagrangian form

$$\psi(x, y) = \varphi(x, y) - \alpha(x) - \beta(y) = -\left(\langle Sx, y\rangle + F(x)\right) \tag{3.5.3}$$

defined on $K \times L = Q \times \mathbb{R}^m$. Now we find

$$\alpha^*(y) = \sup_Q \{\langle -Sx, y\rangle - F(x)\} = \sup_Q \{\langle x, -S^*y\rangle - F(x)\} = F^*(-S^*y)$$

with a domain of definition

$$N = \{y \in \mathbb{R}^m : -S^*y \in Q^*\}.$$

According to Lemma 3.5.1, the vector $z = -S^*y$ for any $y \in \mathbb{R}^m$ is a tension and moreover $\Theta = -S^*(\mathbb{R}^m)$.

From this follows

$$\alpha^*(y) = F^*(-S^*y) = F^*(z)$$

with $z \in Q^* \cap \Theta$.

Furthermore, we have

$$\beta^*(x) = \inf_{\mathbb{R}^m} \langle -Sx, y\rangle = \inf_{\mathbb{R}^m} \langle x, -S^*y\rangle = \inf_{\Theta} \langle x, z\rangle.$$

Since Θ is a linear space, for any $z \in \Theta$ and any real number λ, the element λz is a tension. Consequently $\langle x, \lambda z\rangle = \lambda\langle x, z\rangle$ attains any value, and hence the infimum exists only in case x is a flow because flow and tension are orthogonal to each other. Now

$$\beta^*(x) = \inf_{\Theta} \langle x, z\rangle = 0$$

domain of definition being $M = Q \cap \Phi$. Consequently the dual pair of problems can be formulated as follows:

$$\text{Find } W = \sup_{M} \{\beta^*(x) - \alpha(x)\} = \sup_{Q \cap \Phi} \{-F(x)\}, \tag{P}$$

$$\text{find } W^* = \inf_{N} \{\alpha^*(y) - \beta(y)\} = \inf_{Q^* \cap \Theta} F^*(z). \tag{D}$$

The primal problem (P) is the so called *transportation problem* while the dual problem (D) appears as *potential problem*. In the context of transportation and potential problem, Theorem 2.3.1 has the following special form.

Theorem 3.5.2. *The following three statements are equivalent:*

(i) *There exist points* $(x^0, z^0) \in Q \times \Theta$ *such that for any* $(x, z) \in Q \times \Theta$, *we have*

$$\langle x, z^0\rangle - F(x) \leqq \langle x^0, z^0\rangle - F(x^0) \leqq \langle x^0, z\rangle - F(x^0).$$

(ii) *The sets* $Q \cap \Phi$ *and* $Q^* \cap \Theta$ *are not empty and the problems* (P) *and* (D) *are realizable. Furthermore,*

$$\max_{Q \cap \Phi} \{-F(x)\} = \min_{Q^* \cap \Theta} F^*(z).$$

(iii) *There exist points* $(x^0, z^0) \in Q \cap \Phi \times Q^* \cap \Theta$ *such that*

$$F(x^0) + F^*(z^0) = 0.$$

Of course we can interpret the transportation and potential problem as a special case of convex programming. We just put $f(x) = F(x)$ defined on $C = Q$ and $g^*(z) = 0$ defined on $D^* = \Theta$. Then we get the Lagrangian form

$$\varphi(x, z) = \langle x, z\rangle - F(x)$$

defined on $C \times D^* = Q \times \Theta$. From this we get the conjugate convex functions $F^*(z)$ defined on $C^* = Q^*$ and $g(x) = 0$ defined on $D = \Phi$.

Consequently, we can apply Theorems 3.4.2 and 3.4.3 to the dual pair of problems to get the following theorems of transportation and potential.

Theorem 3.5.3. *If the sets* $Q \cap \Phi$ *and* $Q^* \cap \Theta$ *are not empty the transportation and potential problem is soluble and*

$$\sup_{Q \cap \Phi} \{-F(x)\} = \inf_{Q^* \cap \Theta} F^*(z).$$

Theorem 3.5.4. *If the sets* Q, Φ *and* Q^*, Θ *possesses relative interior points in common, the transportation and potential problem is realizable and we have*

$$\max_{Q \cap \Phi} \{-F(x)\} = \min_{Q^* \cap \Theta} F^*(z).$$

From Theorem 3.5.4, it follows that the statements (i), (ii), (iii) of Theorem 3.5.2 are true in case Q, Φ as well as Q^*, Θ possess relative interior points and of course, according to Theorem 3.5.2, all those statements are equivalent. Because Φ and Θ are subspaces of $\mathbb{R}^n$ any flow $x \in \Phi$ is an interior point of with respect to the topology of Θ. Here the topologies of Φ and Θ are the induced topologies of $\mathbb{R}^n$. It means that any flow x is a relative interior point in Φ and any tension is a relative interior point in Θ. Thus, to satisfy the assumption of Theorem 3.5.4, we have to ensure that there exist a flow $x \in \Phi$ and a tension $z \in \Theta$ such that x is a relative interior point of Q and z is a relative interior point of Q^*. In the following theorems, we find the conditions for existence of such flows and tensions.

Theorem 3.5.5. *Given a parallelepiped* $Q = [b_1, c_1] \times \cdots \times [b_n, c_n]$ *with* $b_i < c_i$.

There exists a flow $x \in \Phi$ *which is an interior point in* Q *if and only if for any elementary co-cycle*

$$\sum_{i\in\omega^-} c_i > \sum_{i\in\omega^+} b_i, \quad \sum_{i\in\omega^+} c_i > \sum_{i\in\omega^-} b_i. \tag{3.5.4}$$

Proof: 1. Let S be the incidence matrix of G. We already know that $x \in \mathbb{R}^n$ is a flow if and only if $Sx = 0$. But it is even sufficient to know that $Sx \geqq 0$ in order to decide whether x is a flow or not. Because, if x is a flow then $Sx = 0$ and $Sx \geqq 0$ is trivial. On the other hand if $Sx \geqq 0$ then $\langle x(\omega(a_i)), x\rangle \geqq 0$ for $i = 1, 2, \ldots, m$. It is not possible that for any i we have $\langle x(\omega(a_i)), x\rangle > 0$. Because, in this case, we have

$$0 < \sum_{i=1}^{m} \sum_{j=1}^{n} \xi_j(\omega(a_i))\, \xi_j = \sum_{j=1}^{n} \xi_j \sum_{i=1}^{m} \xi_j(\omega(a_i)) = \sum_{j=1}^{n} \xi_j(1 - 1) = 0$$

and this is a contradiction. Consequently, $Sx \geqq 0$ really means that $Sx = 0$; thus x is a flow.

2. From the condition (3.5.4), it follows that there exists an $\varepsilon > 0$ not depending on ω such that

$$\begin{aligned} &\sum_{i\in\omega^-} (c_i - \varepsilon) \geqq \sum_{i\in\omega^+} (b_i + \varepsilon);\\ &\sum_{i\in\omega^+} (c_i - \varepsilon) \geqq \sum_{i\in\omega^-} (b_i + \varepsilon). \end{aligned} \tag{3.5.5}$$

Now, if we show that condition (3.5.5) is necessary and sufficient that there exists a flow $x^0 \in \Phi$ such that $b_i + \varepsilon \leqq \xi_i^0 \leqq c_i - \varepsilon$, $i = 1, 2, \ldots, n$, then it follows that $b_i < \xi_i^0 < c_i$ and thus x^0 is an interior point in Q.

3. To prove point 2 we apply solvability condition (3.1.10). In view of point 1, the existence of a flow x^0 which is an interior point in Q is equivalent to the solvability of the system

$$Ax \leqq B$$

with

$$A = \begin{bmatrix} -S \\ -I_n \\ I_n \end{bmatrix}, \quad B = \begin{bmatrix} 0 \\ -\bar{b} \\ \bar{c} \end{bmatrix}, \quad \bar{b} = \begin{bmatrix} b_1 + \varepsilon \\ \vdots \\ b_n + \varepsilon \end{bmatrix}, \quad \bar{c} = \begin{bmatrix} c_1 - \varepsilon \\ \vdots \\ c_n - \varepsilon \end{bmatrix}. \tag{3.5.6}$$

Now, if x is a solution of (3.5.6.), it follows that $Sx \geqq 0$, $\bar{b} \leqq x$, $x \leqq \bar{c}$.

According to the conditions (3.1.10) the system (3.5.6) is solvable if and only if for any $w = (y, u, v) \in \mathbb{R}^{m+} \times \mathbb{R}^{n+} \times \mathbb{R}^{n+}$ with $A^*w = 0$, we have $\langle B, w\rangle \geqq 0$.

4. Next we show that (3.5.5) is equivalent to the condition $\langle B, w\rangle \geqq 0$ for any $w = (y, u, v) \in \mathbb{R}^{m+} \times \mathbb{R}^{n+} \times \mathbb{R}^{n+}$ with $A^*w = 0$. Consequently, (3.5.5) is necessary and sufficient for existence of a solution for (3.5.6) and this proves the theorem. □

Let us consider $w \in \mathbb{R}^{m+} \times \mathbb{R}^{n+} \times \mathbb{R}^{n+}$ with

$$A^*w = -S^*y - u + v = 0 \quad \text{and} \quad \langle \bar{c}, v \rangle \geqq \langle \bar{b}, u \rangle. \tag{3.5.7}$$

In case $y = 0$ we get $u = v$. If we put $u = v = e^i$, it follows from (3.5.7) that

$$b_i < \langle \bar{b}, e^i \rangle = b_i + \varepsilon \leqq \langle \bar{c}, e^i \rangle = c_i - \varepsilon < c_i.$$

But this inequality is true according to the assumption of the theorem. By Lemma 3.5.1 the vector $z = -S^*y$ is a tension. If λ is a real number and $y^\lambda = y + \lambda e = (\eta_1, \eta_2, \ldots, \eta_m) + \lambda(1, \ldots, 1)$, then

$$-S^*y = -S^*y - \lambda S^*e = -S^*y = z$$

i.e. the tension z does not depend on λ but only on y. For this reason, we can drop the assumption $y \in \mathbb{R}^{m+}$, because in case $y \notin \mathbb{R}^{m+}$, we choose λ so great that $y^\lambda \in \mathbb{R}^{m+}$, but the potential y^λ produces the same tension z as y. Now, from (3.5.7) it follows that for any $(u, v) \in \mathbb{R}^{n+} \times \mathbb{R}^{n+}$, with $u - v = z \in \Theta$, we have

$$\langle \bar{c}, v \rangle \geqq \langle \bar{b}, u \rangle.$$

This is especially true for $z = x(\omega)$ where ω is any elementary co-cycle.

Let us define vectors $x(\omega^+)$ and $x(\omega^-)$ as follows:

$$\xi_i(\omega^+) = \begin{cases} 1, & i \in \omega^+, \\ 0, & i \notin \omega^+, \end{cases}$$

$$\xi_i(\omega^-) = \begin{cases} 1, & i \in \omega^-, \\ 0, & i \notin \omega^-. \end{cases}$$

From this follows

$$z = x(\omega) = x(\omega^+) - x(\omega^-).$$

Consequently, $u = x(\omega^+)$ and $v = x(\omega^-)$ is a solution of the equation

$$z = x(\omega) = u - v.$$

Thus,

$$\langle \bar{c}, x(\omega^-) \rangle = \sum_{i \in \omega^-} (c_i - \varepsilon) \geqq \langle \bar{b}, x(\omega^+) \rangle = \sum_{i \in \omega^+} (b_i + \varepsilon).$$

Similar considerations with ω^+ in place of ω^- leads to

$$\sum_{i \in \omega^+} (c_i - \varepsilon) \geqq \sum_{i \in \omega^-} (b_i + \varepsilon).$$

These inequalities together constitute (3.5.5).

Next, under the assumption that the inequalities (3.5.5) are true, we will show that the condition (3.5.7) follows. The system (3.5.5) means that for any co-cycle ω, we have

$$\langle \bar{c}, x(\omega^-) \rangle \geqq \langle \bar{b}, x(\omega^+) \rangle. \tag{3.5.8}$$

This is especially true for the particular case of co-cycle basis. According to Theorem 3.5.1, we can represent any tension $z \in \Theta$ in the form

$$z = \sum_{i=1}^{m-p} \lambda_i x(\omega_i) = \sum_{i=1}^{m-p} \lambda_i x(\omega_i^+) - \sum_{i=1}^{m-p} \lambda_i x(\omega_i^-) = z^+ - z^-.$$

We put $u = z^+$ and $v = z^-$. Then u and v is a solution of $z = u - v$, with $(u, v) \in \mathbb{R}^{n+}$

$\times \mathbb{R}^{n+}$ if and only if $\lambda_i \geqq 0$, $i = 1, 2, \ldots, m - p$. The general form of a solution for $u - v = z \in \Theta$ with $(u, v) \in \mathbb{R}^{n+} \times \mathbb{R}^{n+}$ is $u = z^+ + h$ and $v = z^- + h$ where h is any vector contained in $\mathbb{R}^{n+}$. It is necessary that h is a nonnegative vector, because for any $k = 1, 2, \ldots, n$, either the component $[z^+]_k$ or the component $[z^-]_k$ is zero. Now, according to (3.5.8), we get

$$\begin{aligned}\langle \bar{c}, v\rangle - \langle \bar{b}, u\rangle &= \langle \bar{c}, z^- + h\rangle - \langle \bar{b}, z^+ + h\rangle \\ &= \langle \bar{c}, z^-\rangle - \langle \bar{b}, z^+\rangle + \langle h, \bar{c} - \bar{b}\rangle \geqq \langle \bar{c}, z^-\rangle - \langle \bar{b}, z^+\rangle \\ &= \sum_{i=1}^{m-p} \lambda_i\big(\langle \bar{c}, x(\omega_i^-)\rangle - \langle \bar{b}, x(\omega_i^+)\rangle\big) \geqq 0.\end{aligned}$$

Thus, the condition (3.5.7) is satisfied.

Analogously to Theorem 3.5.5, we have the following theorem concerning tensions.

Theorem 3.5.6. *Given a parallelepiped $Q^* = [b_1^*, C_1^*] \times \cdots \times [b_n^*, C_n^*]$ with $b_i^* < C_i^*$. There exists a tension $z \in \Theta$ which is an interior point in Q^* if and only if for any elementary cycle μ*

$$\sum_{i\in\mu^-} C_i^* > \sum_{i\in\mu^+} b_i^*; \quad \sum_{i\in\mu^+} C_i^* > \sum_{i\in\mu^-} b_i^*. \tag{3.5.9}$$

Proof: The construction of the proof is similar to that of Theorem 3.5.5.

1. The condition (3.5.9) is equivalent to the existence of an $\varepsilon > 0$ which does not depend on cycles μ such that for any cycle μ we have

$$\begin{aligned}\langle \bar{C}^*, x(\mu^-)\rangle &\geqq \langle \bar{b}^*, x(\mu^+)\rangle; \\ \langle \bar{C}^*, x(\mu^+)\rangle &\geqq \langle \bar{b}^*, x(\mu^-)\rangle.\end{aligned} \tag{3.5.10}$$

Here the vectors $x(\mu^+)$ and $x(\mu^-)$ are defined in the same way as for co-cycles; thus

$$x(\mu) = x(\mu^+) - x(\mu^-).$$

Furthermore,

$$\bar{C}^* = \begin{bmatrix} C_1^* - \varepsilon \\ \vdots \\ C_n^* - \varepsilon \end{bmatrix}, \quad \bar{b}^* = \begin{bmatrix} b_1^* + \varepsilon \\ \vdots \\ b_n^* + \varepsilon \end{bmatrix}.$$

We will find out that the condition (3.5.10) is necessary and sufficient for the existence of a tension $z^0 \in \Theta$, with

$$\bar{b}^* \leqq z^0 \leqq \bar{C}^*.$$

Consequently, z^0 is an interior point in Q^*.

2. The existence of a tension which is an interior point in Q^* is equivalent to solvability of the system

$$Ay \leqq B, \quad y \in \mathbb{R}^m \tag{3.5.11}$$

with

$$A = \begin{bmatrix} -S^* \\ S^* \end{bmatrix}, \quad B = \begin{bmatrix} \bar{C}^* \\ -\bar{b}^* \end{bmatrix}$$

i.e. if y is a solution of (3.5.11) then it follows that

$$\bar{b}^* \leqq -S^*y \leqq \bar{C}^*$$

where $-S^*y$ is the corresponding tension for the potential y. According to the solvability condition (3.1.10), there exists a solution for (3.5.11) if and only if

$$\langle B, w\rangle = \langle \bar{C}^*, u\rangle - \langle \bar{b}^*, v\rangle \geqq 0$$

for any $w = (u, v) \in \mathbb{R}^{n+} \times \mathbb{R}^{n+}$, satisfying

$$A^*w = S(v - u) = 0.$$

From $S(v - u) = 0$ follows $v - u \in \Phi$. Now, we can formulate the solvability condition as follows: For any $(u, v) \in \mathbb{R}^{n+} \times \mathbb{R}^{n+}$ with $v - u \in \Phi$, we have

$$\langle \bar{C}^*, u\rangle \geqq \langle \bar{b}^*, v\rangle. \tag{3.5.12}$$

3. We will now show that the conditions (3.5.10) and (3.5.12) are equivalent. Assume (3.5.12) is true and μ is any elementary cycle. Put

$$v = x(\mu^+), \quad u = x(\mu^-).$$

Consequently $v - u = x(\mu^+) - x(\mu^-)$ and $(u, v) \in \mathbb{R}^{n+} \times \mathbb{R}^{n+}$. Thus, according to (3.5.12)

$$\langle \bar{C}^*, x(\mu^-)\rangle \geqq \langle \bar{b}^*, x(\mu^+)\rangle.$$

If we replace μ^+ by μ^- and μ^- by μ^+, we get

$$\langle \bar{C}^*, x(\mu^+)\rangle \geqq \langle \bar{b}^*, x(\mu^-)\rangle.$$

These inequalities together constitutes the condition (3.5.10). Now, we assume that (3.5.10) is true and $u - v \in \Phi$ with $(u, v) \in \mathbb{R}^{n+} \times \mathbb{R}^{n+}$. According to Theorem 3.5.1, we can represent any flow x in the form,

$$x = \sum_{i=1}^{n-m+p} \lambda_i x(\mu_i) = \sum_{i=1}^{n-m+p} \lambda_i x(\mu_i^+) - \sum_{i=1}^{n-m+p} \lambda_i x(\mu_i^-) = x^+ - x^-.$$

Thus, we have $u - v \in \Phi$ if $u = x^+ + h$ and $v = x^- + h$. Here h must be contained in $\mathbb{R}^{n+}$. In fact, either $[x^+]_k = 0$ or $[x^-]_k = 0$, $k = 1, \ldots, n$ and $(u, v) \in \mathbb{R}^n \times \mathbb{R}^{n+}$. Thus the vector h must necessarily be contained in $\mathbb{R}^{n+}$. Furthermore, in case $h = 0$, it follows that $\lambda_i \geqq 0$. Finally, we arrive at the following estimates:

$$\begin{aligned}
\langle \bar{C}^*, u\rangle - \langle \bar{b}^*, v\rangle &= \langle \bar{C}^*, x^+ + h\rangle - \langle \bar{b}^*, x^- + h\rangle \\
&= \langle \bar{C}^*, x^+\rangle - \langle \bar{b}^*, x^-\rangle + \langle h, \bar{C}^* - \bar{b}^*\rangle \\
&= \sum_{i=1}^{n-m+p} \lambda_i\big(\langle \bar{C}^*, x(\mu_i^+)\rangle - \langle \bar{b}^*, x(\mu_i^-)\rangle\big) + \langle h, \bar{C}^* - \bar{b}^*\rangle \geqq 0,
\end{aligned}$$

but this is the condition (3.5.12). □

Theorems 3.5.4, 3.5.5 and 3.5.6 may now be unified to get a more general theorem.

Theorem 3.5.7. *Let $F(x) = \sum_{i=1}^{n} f_i(\xi_i)$ be a convex function defined on Q and $F^*(z) = \sum_{i=1}^{n} f_i^*(\zeta_i)$ the corresponding conjugate convex function defined on Q^*.*

The transportation and potential problem is realizable and the statements (i), (ii), (iii) *of*

Theorem 3.5.2 *are true if for any elementary co-cycle* ω *and for any elementary cycle* μ, *we have*

$$\sum_{i\in\omega^-} c_i > \sum_{i\in\omega^+} b_i, \quad \sum_{i\in\omega^+} c_i > \sum_{i\in\omega^-} b_i;$$

$$\sum_{i\in\mu^-} C_i^* > \sum_{i\in\mu^+} b_i^*, \quad \sum_{i\in\mu^+} C_i^* > \sum_{i\in\mu^-} b_i^*.$$

3.6. Generalized transportation and potential problem in finite directed graphes

To describe concrete practical problems more precisely, Dragan generalized the idea of a flow [7]. He introduced a multicommodity flow in form of a matrix with the property that the lines of the matrix are flows according to the definition given in Section 3.5. In this section we will generalize the idea of flows and tensions by defining them in abstract Banach spaces one may find in works by Walk [35], [36], [37], [38]. Multicommodity flows we will get in form of special cases. Similar to Section 3.5 we construct this theory in the sense of the idea of duality. With the aid of generalized flows and tensions we are also able to describe a dynamic process in form of transportation and potential problems. First we define generalized flows and tensions and investigate relations between them.

Let $G = (X, U)$ be a finite directed graph with $|U| = n$ and $|X| = m$. We assume G to be connected i.e. $p = 1$, and not containing any loops.

Let $E_1, \ldots, E_n$ be Banach spaces and $E_1^*, \ldots, E_n^*$ the corresponding dual spaces containing all linear functionals on E_i, $i = 1, \ldots, n$. We denote by x and x^* elements in form of $x = (x_1, \ldots, x_n) \in E_1 \times \cdots \times E_n = E$ and $x^* = (x_1^*, \ldots, x_n^*) \in E_1^* \times \cdots \times E_n^* = E^*$. Similar to Section 3.5 we associate elements x and x^* with $G = (X, U)$ while we assign to each arc $i \in U$ the element x_i and x_i^* respectively. Now we consider any fixed elements $a \in E$ and $a^* \in E^*$ with $a_i \neq 0$ and $a_i^* \neq 0$ respectively, $i = 1, \ldots, n$.

Definition 3.6.1. We call an element x an *a^*-flow* in G if $a^* \circ x = (\langle a_1^*, x_1\rangle, \ldots, \langle a_n^*, x_n\rangle) \in \Phi$; we call an element x^* an *a-tension* in G if $x^* \circ a = (\langle x_1^*, a_1\rangle, \ldots, \langle x_n^*, a_n\rangle) \in \Theta$.

For fixed elements a^* and a we denote by $\Phi_{a^*} \subseteqq E$ the totality of all a^*-flows and by $\Theta_a \subseteqq E^*$ the totality of all a-tensions in G. Since Φ and Θ are linear spaces it is obvious that Φ_{a^*} and Θ_a are linear spaces. For the special case $E_1 = E_2 = \cdots = E_n = \mathbb{R}$ and $a^* = (1, \ldots, 1)$, $a = (1, \ldots, 1)$ we obviously get $\Phi_{a^*} = \Phi$ and $\Theta_a = \Theta$.

Let us now consider the case $E_1 = \cdots = E_n = E_0$, $a_1^* = \cdots = a_n^* = a_0^* \neq 0$, $a_1 = \cdots = a_n = a_0 \neq 0$. For this special case we denote the linear space of a^*-flows by $\Phi_{a_0^*}$ and the linear space of a-tensions by Θ_{a_0}. A further particularity could be

$$\Phi_0 = \bigcap_{a_0^* \neq 0} \Phi_{a_0^*}, \quad \Theta_0 = \bigcap_{a_0 \neq 0} \Theta_{a_0}. \tag{3.6.1}$$

According to Definition 3.6.1 an element x is contained in Φ_0 if and only if for all linear functionals $a_0^* \in E_0^*$ with $a_0^* \neq 0$ we get

$$\sum_{i\in\omega^+} \langle a_0^*, x_i\rangle - \sum_{i\in\omega^-} \langle a_0^*, x_i\rangle = \left\langle a_0^*, \sum_{i\in\omega^+} x_i - \sum_{i\in\omega^-} x_i\right\rangle = 0. \tag{3.6.2}$$

This equation must be true for any elementary co-cycle $\omega = \omega^+ \cup \omega^-$ of a given connected graph $G(X, U)$. The equation (3.6.1) holds if and only if

$$\sum_{i \in \omega^+} x_i - \sum_{i \in \omega^-} x_i = 0 \tag{3.6.3}$$

for all elementary co-cycles. Similar it is x^* contained in Θ_0 if and only if

$$\sum_{i \in \mu^+} x_i^* - \sum_{i \in \mu^-} x_i^* = 0 \tag{3.6.4}$$

for all elementary cycles μ. Specially for $E_1 = \cdots = E_n = E_0 = \mathbb{R}^r$ we get multicommodity flows investigated by DRAGAN [7]. Moreover besides multicommodity flows according (3.6.3) we may consider multicommodity tensions. The equations (3.6.3) and (3.6.4) show that a matrix of order (k, n) is a multicommodity flow or a multicommodity tension then and only then if their lines are contained in Φ or Θ respectively.

Next we investigate a possible representation of generalized flows and tensions. We choose $a \in E$ and $a^* \in E^*$ such that $a^* \circ a = (\langle a_1^*, a_1 \rangle, \ldots, \langle a_n^*, a_n \rangle) = (1, \ldots, 1)$. Let μ be any cycle and ω be any co-cycle of the graph. We now identify cycles and co-cycles with corresponding vectors

$$\mu_i = \begin{cases} 1, & \text{if } i \in \mu^+ \\ -1, & \text{if } i \in \mu^-, \\ 0, & \text{if } i \notin \mu \end{cases} \qquad \omega_i = \begin{cases} 1, & \text{if } i \in \omega^+ \\ -1, & \text{if } i \in \omega^- \\ 0, & \text{if } i \notin \omega \end{cases}$$

contained in $\mathbb{R}^n$. Since $\langle a_i^*, a_i \rangle = 1$, $i = 1, \ldots, n$ we obviously get

$$\mu \circ a = (\mu_1 a_1, \ldots, \mu_n a_n) \in \Phi_{a^*}$$
$$a^* \circ \omega = (\omega_1 a_1^*, \ldots, \omega_n a_n^*) \in \Theta_a.$$

More general for $x \in E$ and a cycle μ of G the element $\mu \circ x \in E$ will be a a^*-flow if and only if $\langle a_i^*, x_i \rangle$ is constant for all i with $\mu_i \neq 0$. Analogously $x^* \circ \omega$ will be a a-tension if and only if $\langle x_i^*, a_i \rangle$ is constant for all i with $\omega_i \neq 0$. Now we consider a cycle basis $\mu^1, \ldots, \mu^k$ and a co-cycle basis $\omega^1, \ldots, \omega^l$. Since Φ_{a^*} and Θ_a are linear spaces it is obvious that

$$L_{a^*} = \left\{ x \in E : x = \sum_{s=1}^{k} \lambda_s \mu^s \circ a^s, \quad a^s \in E, \langle a_i^*, a_i^s \rangle = \varrho_s \neq 0 \right.$$
$$\left. \text{for all } i \text{ with } \mu_i^s \neq 0 \right\}$$

$$L_a = \left\{ x^* \in E^* : x^* = \sum_{s=1}^{l} \sigma_s a^{*s} \circ \omega^s, a^{*s} \in E^*, \langle a_i^{*s}, a_i \rangle = \sigma_s \neq 0 \right.$$
$$\left. \text{for all } i \text{ with } \omega_i^s \neq 0 \right\}$$

are linear subspaces of Φ_{a^*} and Θ_a respectively. We denote

$$N_{a_i^*} = \{x_i \in E_i : \langle a_i^*, x_i \rangle = 0\}, \quad N_{a_i} = \{x_i^* \in E_i^* : \langle x_i^*, a_i \rangle = 0\}$$

and consider

$$N_a^* = N_{a_1^*} \times \cdots \times N_{a_n^*}, \quad N_a = N_{a_1} \times \cdots \times N_{a_n}.$$

Obviously N_{a^*} and N_a are also subspaces of Φ_{a^*} and Θ_a respectively thus

$$L_{a^*} + N_{a^*} \subseteqq \Phi_{a^*}, \quad L_a + N_a \subseteqq \Theta_a. \tag{3.6.5}$$

Theorem 3.6.1. *The spaces Φ_{a*} and Θ_a of a^*-flows and a tensions can be represented in form of*

$$\Phi_{a*} = L_{a*} \oplus N_{a*}, \quad \Theta_a = L_a \oplus N_a .$$

Here the notation $\oplus$ has the indication that the elements $0 \in \Phi_{a*}$ and $0 \in \Theta_a$ can be represented only in a trivial way. The symbol 0 always denotes the zero-element of the corresponding space.

Proof: It is sufficient to prove the representation of Φ_{a*} because the realization of the proof for Θ_a will be analogous. Let x be an element contained in Φ_{a*} thus $a^* \circ x \in \Phi$. According to the representation of Φ given in Section 3.5 there exist uniquely defined real numbers $\lambda_1, \ldots, \lambda_k$ such that

$$a^* \circ x = \sum_{s=1}^{k} \lambda_s \mu^s .$$

Now we choose k elements $a^1, \ldots, a^k \in E$ with

$$\langle a_i^*, a_i^s \rangle = \varrho_s \neq 0 \quad \text{for all } i \text{ with } \mu_i^s \neq 0 .$$

It follows

$$\sum_{s=1}^{k} \frac{\lambda_s}{\varrho_s} \mu^s \circ a^s \in L_{a*}$$

with

$$a^* \circ \sum_{s=1}^{k} \frac{\lambda_s}{\varrho_s} \mu^s \circ a^s = \sum_{s=1}^{k} \lambda_s \mu^s .$$

Consequently

$$a^* \circ \left(x - \sum_{s=1}^{k} \frac{\lambda_s}{\varrho_s} \mu^s \circ a^s \right) = 0$$

and this implies

$$x - \sum_{s=1}^{k} \frac{\lambda_s}{\varrho_s} \mu^s \circ a^s \in N_{a*}$$

thus $x \in L_{a*} + N_{a*}$ i.e. $\Phi_{a*} \subseteqq L_{a*} + N_{a*}$. According to (3.6.5) now it follows $\Phi_{a*} = L_{a*} + N_{a*}$. To complete the proof we have to show that the zero-element can be represented trivial only. Really from

$$0 = \sum_{s=1}^{k} \lambda_s \mu^s \circ a^s + x^0 \quad \text{with} \quad x^0 \in L_{a*}$$

it follows

$$a^* \circ 0 = 0 = \sum_{s=1}^{k} \lambda_s \varrho_s \mu^s .$$

Since $\mu^1, \ldots, \mu^k$ is a cycle basis it follows $\lambda_1 \varrho_1 = 0, \ldots, \lambda_k \varrho_k = 0$. From $\varrho_s \neq 0$ we get $\lambda_1 = \cdots = \lambda_k = 0$. Consequently $x^0 = 0$. □

The property that Φ and Θ are orthogonal complements of $\mathbb{R}^n$ will not remain for the generalized case. If we define a scalar product in $E^* \times E$ in form of $\langle x^*, x \rangle = \sum_{i=1}^{n} \langle x_i^*, x_i \rangle$

one can obviously see that for $(x^*, x) \in \Theta_a \times \Phi_{a^*}$ generally we have $\langle x^*, x\rangle \neq 0$. From this point of view we like to find at least subspaces of Φ_{a^*} and Θ_a such that the subspaces are orthogonal to each other. This will be an important matter for constructing such Lagrangian forms which produce a dual pair of problems.

Let $(a^*, a) \in E^* \times E$ be any elements with $\langle a_i^*, a_i\rangle = 1$, $i = 1, \ldots, n$. We use the following notations:

$$L_{a^*,a} = \left\{x \in E : x = \sum_{s=1}^{k} \lambda_s \mu^s \circ a\right\}, \quad \Phi_{a^*,a} = L_{a^*,a} \oplus N_{a^*}$$
$$L_{a,a^*} = \left\{x^* \in E^* : x^* = \sum_{s=1}^{l} \sigma_s a^* \circ \omega^s\right\}, \quad \Theta_{a,a^*} = L_{a,a^*} \oplus N_a . \tag{3.6.6}$$

Obviously $\Phi_{a^*,a}$ and Θ_{a,a^*} are subspaces of Φ_{a^*} and Θ_a respectively.

Theorem 3.6.2. *Let $G = (X, U)$ be a graph with the property that each arc of G is contained in at least one cycle of G. If $a^*, b^* \in E^*$ and $a, b \in E$ are any elements with $\langle a_i^*, a_i\rangle = 1$, $\langle b_i^*, b_i\rangle = 1$, $i = 1, \ldots, n$, then $\Phi_{a^*,a} = \Phi_{b^*,b}$ and $\Theta_{a,a^*} = \Theta_{b,b^*}$ if and only if there exists a real number $\lambda \neq 0$ such that $b^* = \lambda a^*$ and $b = \lambda a$ respectively.*

Proof: We prove this theorem for flows for the same reason as in Theorem 3.6.1.

Let $b^* = \lambda a^*$ and $x \in \Phi_{a^*,a}$ thus $x = \sum_{s=1}^{k} \lambda_s \mu^s \circ a + x^0$, $x^0 \in N_{a^*}$. Then we get

$$b^* \circ x = (\lambda a^*) \circ x = \lambda(a^* \circ x) \in \Phi$$

i.e. x is a b^*-flow. Furthermore from

$$\langle b_i^*, b_i\rangle = 1 = \langle a_i^*, a_i\rangle = \langle \lambda a_i^*, b_i\rangle$$

it follows $a_i - \lambda b_i = \bar{x}_i^0 \in N_{a_i^*}$. If we substitute $a_i = \lambda b_i + \bar{x}_i^0$ in the representation of x we get

$$x = \sum_{s=1}^{k} \lambda_s \mu^s \circ (\lambda b + \bar{x}^0) = \sum_{s=1}^{k} \lambda \lambda_s \mu^s \circ b + \sum_{s=1}^{k} \lambda_s \mu^s \circ \bar{x}^0 + x^0 .$$

Since $b^* = \lambda a^*$ it follows

$$b^* \circ \left(\sum_{s=1}^{k} \lambda_s \mu^s \circ \bar{x}^0 + x^0\right) = 0$$

hence $\sum_{s=1}^{k} \lambda_s \mu^s \circ \bar{x}^0 + x^0 \in N_{a^*}$ thus $x \in \Phi_{b^*,b}$. Reversely from $x \in \Phi_{b^*,b}$ we get $x \in \Phi_{a^*,a}$ thus $\Phi_{a^*,a} = \Phi_{b^*,b}$. We now assume $\Phi_{a^*,a} = \Phi_{b^*,b}$. For any cycle μ valid $\mu \circ a \in \Phi_{a^*,a} = \Phi_{b^*,b}$. Consequently

$$b^* \circ (\mu \circ a) = (\mu_1 \langle b_1^*, a_1\rangle, \ldots, \mu_n \langle b_n^*, a_n\rangle) \in \Phi . \tag{3.6.7}$$

Because any arc of G is contained in at least one cycle, (3.6.7) is possible only if $\langle b_i^*, a_i\rangle = \lambda$ for $i = 1, \ldots, n$. We will now prove $N_{a^*} = N_{b^*}$. Let x^0 be an element in N_{a^*}. The assumption $\Phi_{a^*,a} = \Phi_{b^*,b}$ implies $b^* \circ x^0 \in \Phi$. Moreover for any real numbers $\sigma_1, \ldots, \sigma_n$ unequal to zero it is $(\sigma_1 x_1^0, \ldots, \sigma_n x_n^0) \in N_{a^*}$ thus

$$b^* \circ (\sigma_1 x_1^0, \ldots, \sigma_n x_n^0) = (\sigma_1 \langle b_1^*, x_1^0\rangle, \ldots, \sigma_n \langle b_n^*, x_n^0\rangle) \in \Phi .$$

But this is possible only for $\langle b_i^*, x_i^0\rangle = 0$, $i = 1, \ldots, n$ i.e. $x^0 \in N_b^*$. Analogously we get $N_{b^*} \subseteq N_{a^*}$ and finally $N_{a^*} = N_{b^*}$. From $\langle b_i^*, a_i\rangle = \lambda$, $i = 1, \ldots, n$, and $N_{a^*} = N_{b^*}$

it follows $\lambda \neq 0$. If we consider any element $x \in \Phi_{a^*,a} = \Phi_{b^*,b}$ and

$$x = \sum_{s=1}^{k} \lambda_s \mu^s \circ a + x^0, \quad x^0 \in N_{a^*}$$

then we get

$$b^* \circ x = \sum_{s=1}^{k} \lambda_s b^* \circ (\mu^s \circ a) + b^* \circ x^0 = \sum_{s=1}^{k} \lambda \lambda_s \mu^s + \lambda a^* \circ x^0$$

$$= (\lambda a^*) \circ \left(\sum_{s=1}^{k} \lambda_s \mu^s \circ a + x^0 \right) = (\lambda a^*) \circ x$$

and consequently $b^* = \lambda a^*$ for all $x \in \Phi_{a^*,a} = \Phi_{b^*,b}$ which is the assertion of the theorem. □

Corollary 3.6.1. *By Theorem 3.6.2 it follows from* $x \in \Phi_{a^*,a} = \Phi_{b^*,b}$ *and*

$$x = \sum_{s=1}^{k} \lambda_s \mu^s \circ a + x^0 = \sum_{s=1}^{k} \bar{\lambda}_s \mu^s \circ b + \bar{x}^0$$

the relation $\bar{\lambda}_s = \lambda \lambda_s$, $s = 1, \ldots, k$.

This is a consequence of the fact that $\mu^1, \ldots, \mu^k$ are linearly independent.

Now we calculate $\langle x^*, x \rangle = \sum_{i=1}^{n} \langle x_i^*, x_i \rangle$ for $(x^*, x) \in \Theta_{a,a^*} \times \Phi_{a^*,a}$:
According to (3.6.5) we have

$$x = \sum_{s=1}^{k} \lambda_s \mu^s \circ a + x^0, \quad x^* = \sum_{t=1}^{l} \sigma_t a^* \circ \omega^t + x^{*0}.$$

Hence

$$\langle x^*, x \rangle = \langle x^{*0}, x^0 \rangle + \left\langle x^{*0}, \sum_{s=1}^{k} \lambda_s \mu^s \circ a \right\rangle + \left\langle \sum_{t=1}^{l} \sigma_t a^* \circ \omega^t, x^0 \right\rangle$$

$$+ \left\langle \sum_{t=1}^{l} \sigma_t a^* \circ \omega^t, \sum_{s=1}^{k} \lambda_s \mu^s \circ a \right\rangle.$$

Since $\langle x_i^{*0}, a_i \rangle = 0$, $\langle a_i^*, x_i^0 \rangle = 0$, $\langle \omega^t, \mu^s \rangle = 0$, $i = 1, \ldots, n$; $t = 1, \ldots, l$; $s = 1, \ldots, k$ it follows $\langle x^*, x \rangle = \langle x^{*0}, x^0 \rangle$. Thus also the subspaces $\Phi_{a^*,a}$ and Θ_{a,a^*} are not orthogonal to each other. If we in particular consider $\Phi_0 = \bigcap_{a_0 \neq 0} \Phi_{a_0^*}$, $\Theta_0 = \bigcap_{a_0 \neq 0} \Theta_{a_0}$ we will find that Φ_0 and Θ_0 are orthogonal. Namely by Theorem 3.6.1 we get for $\Phi_{a_0^*}$ and Θ_{a_0} the following representation:

$$\Phi_{a_0^*} = \left\{ x \in E_0 \times \cdots \times E_0 : x = \sum_{s=1}^{k} \lambda_s \mu^s \circ a^s; \, a_1^s = \cdots = a_n^s = a_0^s \neq 0 \right\}$$
$$\oplus N_{a_0^*} \times \cdots \times N_{a_0^*} \tag{3.6.8}$$
$$\Theta_{a_0} = \left\{ x^* \in E_0^* \times \cdots \times E_0^* : x^* = \sum_{s=1}^{l} \sigma_s a^{*s} \circ \omega^s, \, a_1^{*s} = \cdots = a_n^{*s} = a_0^{*s} \neq 0 \right\}$$
$$\oplus N_{a_0} \times \cdots \times N_{a_0}.$$

From (3.6.8) we get a representation for Φ_0 and Θ_0 in form of

$$\Phi_0 = \left\{ x \in E_0 \times \cdots \times E_0 : x = \sum_{s=1}^{k} \lambda_s \mu^s \circ a^s, \, a_1^s = \cdots = a_n^s = a_0^s \neq 0 \right\},$$
$$\Theta_0 = \left\{ x^* \in E_0^* \times \cdots \times E_0^* : x^* = \sum_{s=1}^{l} \sigma_s a^{*s} \circ \omega^s, \, a_1^{*s} = \cdots = a_n^{*s} = a_0^{*s} \neq 0 \right\}. \tag{3.6.9}$$

and consequently for all $(x, x^*) \in \Phi_0 \times \Theta_0$:

$$\langle x^*, x\rangle = \sum_{s=1}^{k} \sum_{t=1}^{l} \lambda_s \sigma_t \langle a_0{}^{*t}, a_0{}^s\rangle \langle \omega^t, \mu^s\rangle = 0 .$$

If we in particular choose $E_0 = \mathbb{R}^r$ we get the special case of multicommodity flows thus multicommodity flows and multicommodity tensions are orthogonal to each other.

Now we investigate correlations between the incident matrix S of the graph G and generalized flows and tensions. If we enumerate all vertices of G from 1 to m such that $X = \{r_1, \ldots, r_m\}$, then we denote $x(\omega(r_1)) = \overline{\omega}^1, \ldots, x(\omega(r_m)) = \overline{\omega}^m$. Hence the matrix

$$S = \begin{bmatrix} \overline{\omega}^1 \\ \vdots \\ \overline{\omega}^m \end{bmatrix}$$

is the incident matrix of the graph G of order (m, n). Beside this we introduce a (m, n)-matrix in form of

$$S(a^*) = \begin{bmatrix} a^* \circ \overline{\omega}^1 \\ \vdots \\ a^* \circ \overline{\omega}^m \end{bmatrix} = \begin{bmatrix} \overline{\omega}_1{}^1 a_1{}^* & \cdots & \overline{\omega}_n{}^1 a_n{}^* \\ \vdots & & \vdots \\ \overline{\omega}_1{}^m a_1{}^* & \cdots & \overline{\omega}_n{}^m a_n{}^* \end{bmatrix}.$$

Theorem 3.6.3. *x is an element contained in $\Phi_{a^*,a}$ if and only if $s(a^*)\, x = 0$.*

Proof: First we assume $x \in \Phi_{a^*,a}$. According to (3.6.6) we get

$$x = \sum_{s=1}^{k} \lambda_s \mu^s \circ a + x^0, \quad x^0 \in N_{a^*}.$$

We calculate

$$S(a^*)\, x^0 = \begin{bmatrix} \overline{\omega}_1{}^1 a_1{}^* & \cdots & \overline{\omega}_n{}^1 a_n{}^* \\ \vdots & & \vdots \\ \overline{\omega}_1{}^m a_1{}^* & \cdots & \overline{\omega}_n{}^m a_n{}^* \end{bmatrix} \begin{bmatrix} x_1{}^0 \\ \vdots \\ x_n{}^0 \end{bmatrix} = \begin{bmatrix} \sum_{i=1}^{n} \overline{\omega}_i{}^1 \langle a_i{}^*, x_i{}^0\rangle \\ \vdots \\ \sum_{i=1}^{n} \overline{\omega}_i{}^m \langle a_i{}^*, x_i{}^0\rangle \end{bmatrix},$$

$$S(a^*) \left[\sum_{s=1}^{k} \lambda_s \mu^s \circ a\right] = \begin{bmatrix} \overline{\omega}_1{}^1 a_1{}^* & \cdots & \overline{\omega}_n{}^1 a_n{}^* \\ \vdots & & \vdots \\ \overline{\omega}_1{}^m a_1{}^* & \cdots & \overline{\omega}_n{}^m a_n{}^* \end{bmatrix} \begin{bmatrix} \mu_1{}^s a_1 \\ \vdots \\ \mu_n{}^s a_n \end{bmatrix} \begin{bmatrix} \sum_{s=1}^{k} \lambda_s \langle \overline{\omega}^1, \mu^s\rangle \\ \vdots \\ \sum_{s=1}^{k} \lambda_s \langle \overline{\omega}^m, \mu^s\rangle \end{bmatrix}.$$

Since $x_i{}^0 \in N_{a_i*}$ and $\langle \overline{\omega}^j, \mu^s\rangle = 0$, $j = 1, \ldots, m$; $s = 1, \ldots, k$, it follows $S(a^*) = 0$.

Now let $x \in E$ such that $S(a^*)\, x = 0$. This equation implies $\langle \overline{\omega}^j, a^* \circ x\rangle = 0$ for $j = 1, \ldots, m$. Since Φ and Θ are orthogonal to each other and $\overline{\omega}^j \in \Theta$ it follows $a^* \circ x \in \Phi$ thus there exist real numbers $\lambda_1, \ldots, \lambda_k$ such that $a^* \circ x = \sum_{s=1}^{k} \lambda_s \mu^s$. Consequently we get

$$a^* \circ \left(x - \sum_{s=1}^{k} \lambda_s \mu^s \circ a\right) = 0$$

thus $x - \sum_{s=1}^{k} \lambda_s \mu^s \circ a = x^0 \in N_{a^*}$ which proves the assertion of the theorem. □

To characterize generalized tensions by potentials we must reduce our considerations on the subspace Θ_{a_0}. We assign to each vertex r_j the component y_j^* of $y^* \in E_0^* \times \cdots \times E_0^* = (E_0^*)^m$.

Definition 3.6.2. An element $y^* \in (E_0^*)^m$ is said to be a *potential* of an element $x^* \in (E_0^*)^n$ if for any arc $i = (r_{j_1}, r_{j_2}) \in U$ holds $x_i^* = y_{r_{j_2}}^* - y_{r_{j_1}}^*$.

Theorem 3.6.4. *$x^* \in (E_0^*)^n$ is an element contained in Θ_{a_0} if and only if there exists a corresponding potential y^* of x^*. The potential y^* is uniquly defined up to an additional element $y^{*0} \in (N_{a_0})^m$.*

Proof: Let $x^* \in \Theta_{a_0}$ and r_{i_0} any vertex of G. We assign to r_{i_0} any linear functional $y_{r_{i_0}}^* \in E_0^*$ and assume without loss of generality $y_{r_{i_0}}^* = 0$. For construction of a corresponding potential we now apply the following algorithm: If we consider any arc $i \in U$ we put

$$y_{r_{i_2}}^* = \begin{cases} y_{r_{i_1}}^* + x_i^* & \text{if } i = (r_{i_1}, r_{i_2}), \\ y_{r_{i_1}}^* - x_i^* & \text{if } i = (r_{i_2}, r_{i_1}). \end{cases} \tag{3.6.10}$$

By this algorithm we assign to all vertices r_j, $j = 1, \ldots, m$, a linear functional $y_{r_j}^* \in E_0^*$. We now show that $y_{r_j}^*$ generated by (3.6.10) is uniquly defined up to an additional element $y_{r_j}^{*0} \in N_{a_0}$. Namely if Π_1 and Π_2 are two series of consecutive arcs, i.e. two chains from r_{i_0} to r_i then we represent Π_1 and Π_2 in form of $\Pi_j = (\sigma_1^j, \ldots, \sigma_n^j)$, $j = 1, 2$ with,

$$\sigma_i^j = \begin{cases} 1, & i \in \Pi_j^+, \\ -1, & i \in \Pi_j^-, \\ 0, & i \notin \Pi_j. \end{cases}$$

Here by Π_j^+ we denote the set of all arcs we pass in the direction of their orientation and by Π_j^- we denote the set of all arcs we pass in the opposite direction of their orientation. Obviously $\Pi_1 - \Pi_2$ is a cycle in G if $-$ means to pass Π_2 in opposite direction. By algorithm (3.6.10) we get

$$\langle \Pi_1, x^* \rangle = \sum_{i=1}^{n} \sigma_i^1 x_i^* = y_{r_i}^*,$$

$$\langle \Pi_2, x^* \rangle = \sum_{i=1}^{n} \sigma_i^2 x_i^* = \bar{y}_{r_i}^*.$$

In case $\bar{y}_{r_i}^*(a_0) \neq y_{r_i}^*(a_0)$ it follows

$$\sum_{i=1}^{n} (\sigma_i^1 - \sigma_i^2)\, x_i^*(a_0) = \langle \Pi_1 - \Pi_2, x^* \circ a_0 \rangle = y_{r_i}^*(a_0) - \bar{y}_{r_i}^*(a_0) \neq 0.$$

Since $\Pi_1 - \Pi_2$ is a cycle there exists an elementary cycle μ such that $\langle \mu, x^* \circ a_0 \rangle \neq 0$ thus $x^* \circ a_0 \notin \Theta$ and consequently $x^* \notin \Theta_{a_0}$. Then we get $(y_{r_i}^* - \bar{y}_{r_i}^*)(a_0) = 0$ thus $y_{r_i}^* - \bar{y}_{r_i}^* = y_{r_i}^{*0} \in N_{a_0}$. It now follows that $y^* \in (E_0^*)^m$ constructed by (3.6.9) is a potential to x^* uniquly defined up to an additional element $y^{*0} \in (N_{a_0})^m$.

To prove sufficiently, we assume the existence of a potential y^* to x^*. Let μ be any elementary cycle of G. Let $(i_1, \ldots, i_\varrho)$ be the order of arcs by passing through μ and $(r_{j_1}, \ldots, r_{j_\varrho}, r_{j_1})$ the order of corresponding vertices. Then we get

$$\mu_{i_1} x_{i_1}^*(a_0) = y_{j_2}^*(a_0) - y_{j_1}^*(a_0)$$

$$\mu_{i_2} x_{i_2}^*(a_0) = y_{j_3}^*(a_0) - y_{j_2}^*(a_0)$$

$$\vdots$$

$$\mu_{i_\varrho} x_{i_\varrho}^*(a_0) = y_{j_1}^*(a_0) - y_{j_\varrho}^*(a_0).$$

If we sum this equations we get

$$\sum_{i=1}^{n} \mu_i x_i^*(a_0) = 0$$

consequently $(x_1^*(a_0), \ldots, x_n^*(a_0)) \in \Theta$ thus $x^* \in \Theta_{a_0}$.

Corollary 3.6.2. *Let S be the incident matrix of G. It is x^* an element contained in Θ_{a_0} if and only if there exists $y^* \in (E_0^*)^m$ such that $x^* = -S^*(y^*)$.*

This follows immediately from Theorem 3.6.4.

Corollary 3.6.3. *We denote by $\Theta_{a_0}/(N_{a_0})^n$ and $(E_0^*)^m/(N_{a_0})^m$ the factor spaces of Θ_{a_0} and $(E_0^*)^m$ respectively. These factor spaces are isomorphically.*

Proof: Let $\{z^*\} \in \Theta_{a_0}/(N_{a_0})^n$ be a class of equivalent elements. Furthermore let $x^* = z^* + x^{*0}$, $x^{*0} \in (N_{a_0})^n$ be any representative of this class. According Theorem 3.6.4 there exists a corresponding potential $w^* \in (E_0^*)^m$ of z^*. Now we show that a corresponding potential y^* of x^* is contained in the class of equivalent elements $\{w^*\}$. For that we assume w^* to be generated by application of algorithm (3.6.10), i.e. we consider any two vertices $r_0, r_j \in X$ and put $w_{r_0}^* \equiv 0$ and $\langle \Pi_{r_j}, z^* \rangle = w_{r_j}^*$ where Π_{r_j} denotes a chain from r_0 to r_j. The vertices you pass by traversing through Π_{r_j} may be given in form of $(r_0, r_1, \ldots, r_j)$. Then we get $w_{r_{j-1}}^* = \langle \Pi_{r_{j-1}}, z^* \rangle$ and $\Pi_{r_{j-1}}$ denotes the chain from r_0 to r_{j-1}. Now we put

$$\begin{aligned} y_{r_j}^* &= w_{r_j}^* + \langle \Pi_{r_j}, x^{*0} \rangle, \\ y_{r_{j-1}}^* &= w_{r_{j-1}}^* + \langle \Pi_{r_{j-1}}, x^{*0} \rangle. \end{aligned} \tag{3.6.11}$$

It is obviously $\langle \Pi_{r_j}, x^{*0} \rangle$ and $\langle \Pi_{r_{j-1}}, x^{*0} \rangle$ contained in N_{a_0} consequently the element $y^* \in (E_0^*)^m$ constructed by (3.6.11) is contained in the class $\{w^*\}$. Moreover if we consider the arc $i = (r_{j-1}, r_j)$ it follows

$$y_{r_j}^* - y_{r_{j-1}}^* = w_{r_j}^* - w_{r_{j-1}}^* + \langle \Pi_{r_j} - \Pi_{r_{j-1}}, x^{*0} \rangle = z_i^* + x_i^{*0} = x_i^*$$

thus y^* is a potential of x^*. If we assign the classes $\{z^*\} \leftrightarrow \{w^*\}$ to each other it follows the assertion of Corollary 3.6.3. □

For generalized transportation and potential problems we have to consider generalized flows and tensions with constrains. In the following theorems we find conditions for existence of such flows and tensions.

Theorem 3.6.5. *Let $C_i \subseteq E_i$, $i = 1, \ldots, n$, be convex and compact sets. There exists a a^*-flow $x \in \Phi_{a^*}$ with $x_i \in C_i$ if and only if for any elementary co-cycle ω*

$$0 \in \sum_{i \in \omega^+} a_i^*(C_i) - \sum_{i \in \omega^-} a_i^*(C_i). \tag{3.6.12}$$

Proof: Since C_i are compact it follows $a_i^*(C_i) = [\alpha_{i_1}, \alpha_{i_2}]$ are finite closed intervals. Let $x \in \Phi_{a^*}$ and $x_i \in C_i$ thus $a_i^*(x_i) \in [\alpha_{i_1}, \alpha_{i_2}]$. We consider any elementary co-cycle ω. Since $a^* \circ x \in \Phi$ and the spaces Φ and Θ are orthogonal complements of $\mathbb{R}^n$ we get

$$0 = \langle \omega, a^* \circ x \rangle = \sum_{i \in \omega^+} a_i^*(x_i) - \sum_{i \in \omega^-} a_i^*(x_i) \leqq \sum_{i \in \omega^+} \alpha_{i_2} - \sum_{i \in \omega^-} \alpha_{i_1},$$

$$0 = \langle -\omega, a^* \circ x \rangle = \sum_{i \in \omega^-} a_i^*(x_i) - \sum_{i \in \omega^+} a_i^*(x_i) \leqq \sum_{i \in \omega^-} \alpha_{i_2} - \sum_{i \in \omega^+} \alpha_{i_1}$$

Consequently it holds

$$\sum_{i\in\omega^+} \alpha_{i_1} - \sum_{i\in\omega^-} \alpha_{i_2} \leqq 0 \leqq \sum_{i\in\omega^+} \alpha_{i_2} - \sum_{i\in\omega^-} \alpha_{i_1}$$

and this is equivalent to

$$0 \in \left[\sum_{i\in\omega^+} \alpha_{i_1} - \sum_{i\in\omega^-} \alpha_{i_2}, \sum_{i\in\omega^+} \alpha_{i_2} - \sum_{i\in\omega^-} \alpha_{i_1}\right] = \left[\sum_{i\in\omega^+} \alpha_{i_1}, \sum_{i\in\omega^+} \alpha_{i_2}\right] - \left[\sum_{i\in\omega^-} \alpha_{i_1}, \sum_{i\in\omega^-} \alpha_{i_2}\right]$$
$$= \sum_{i\in\omega^+} a_i^*(C_i) - \sum_{i\in\omega^-} a_i^*(C_i).$$

To prove sufficiently we assume

$$0 \in \sum_{i\in\omega^+} a_i^*(C_i) - \sum_{i\in\omega^-} a_i^*(C_i).$$

By a theorem from HOFFMANN [4] there exists an element $\varphi \in \Phi$ with $\varphi_i \in a_i^*(C_i)$. Since a_i^* are linear continuous functionals there exist points $x_i \in C_i$ such that $a_i^*(x_i) = \varphi_i$ hence $x = (x_1, \ldots, x_n) \in (c_i)^n$ with $a^* \circ x \in \Phi$ thus $x \in \Phi_{a^*}$.

Analogously you can prove

Theorem 3.6.6. *Let $C_i^* \subseteqq E_i^*$, $i = 1, \ldots, n$, be convex and compact sets. There exists an a-tension $x^* \in \Theta_a$ with $x_i^* \in C_i^*$ if and only if for any elementary cycle μ*

$$0 \in \sum_{i\in\mu^+} C_i^*(a_i) - \sum_{i\in\mu^-} C_i^*(a_i). \tag{3.6.13}$$

Finally let us consider the special case of multicommodity flows. We get the representation of multicommodity flows from (3.6.9) for $E_0 = \mathbb{R}^r$ thus

$$\Phi_0 = \left\{\varphi \in (\mathbb{R}^r)^n\colon\ \varphi = \sum_{s=1}^{k} \lambda_s \mu^s \circ a^s,\ a_1^s = \cdots = a_n^s = a_0^s \neq 0\right\}.$$

From this representation you can see immediately that a matrix

$$\varphi = (\varphi^1, \ldots, \varphi^n) = \begin{bmatrix} \varphi_{11} & \cdots & \varphi_{1n} \\ \vdots & & \vdots \\ \varphi_{r1} & \cdots & \varphi_{rn} \end{bmatrix}$$

is a multicommodity flow if and only if $(\varphi_{j1}, \ldots, \varphi_{jn}) \in \Phi$, $j = 1, \ldots, r$. Let $C_1, \ldots, C_n$ be squares in $\mathbb{R}^r$ and $\varphi \in \Phi_0$ with $\varphi^i \in C_i$. Then it follows

$$\langle e_j, \varphi^i\rangle = \varphi_{ji} \in \langle e_j, C_i\rangle \quad \text{with} \quad e_j \in \mathbb{R}^r,\ e_j = (0, \ldots, 1, 0, \ldots, 0),$$

$j = 1, \ldots, r$. Consequently $(\varphi_{j1}, \ldots, \varphi_{jn}) \in \langle e_j, C_1\rangle \times \cdots \times \langle e_j, C_n\rangle$. According to Theorem 3.6.5 we get for any elementary co-cycle

$$0 \in \sum_{i\in\omega^+} \langle e_j, C_i\rangle - \sum_{i\in\omega^-} \langle e_j, C_i\rangle = \left\langle e_j, \sum_{i\in\omega^+} C_i - \sum_{i\in\omega^-} C_i \right\rangle, \tag{3.6.14}$$

$j = 1, \ldots, r$. Thus in the special case of multicommodity flows the condition (3.6.12) appears in form of (3.6.14). Analogous considerations are possible for multicommodity tensions.

To produce a pair of generalized transportation and potential problems in finite directed graphs we have to construct a corresponding Lagrangian form defined on suitable spaces of generalized flows and tensions. For that we have to consider flows and tensions which are orthogonal to each other. We denote

$$[\Phi_{a*}]^* = \{x^* \in \Theta_a : \langle x^*, z\rangle = 0 \quad \text{for all } z \in \Phi_{a*}\}.$$

The following two statements are obviously equivalent:

(i) $\langle x^*, z\rangle = 0$ for all $z \in \Phi_{a*}$;
(ii) $\langle x^*, x\rangle = 0$ for all $x \in L_{a*}$ and $\langle x^*, x^0\rangle = 0$ for all $x^0 \in N_{a*}$.

From $\langle x^*, x^0\rangle = 0$ for all $x^0 \in N_{a*}$ it follows $\langle x_i^*, x_i^0\rangle = 0$ for all $x_i^0 \in N_{a_i^*}$, $i = 1, \ldots, n$. Consequently

$$x_i^* = \alpha_i a_i^*, \quad \alpha_i \in \mathbb{R}, \quad i = 1, \ldots, n. \tag{3.6.15}$$

We now calculate $\langle x^*, x\rangle = 0$ and apply (3.6.15). It is

$$\langle x^*, x\rangle = \left\langle x^*, \sum_{s=1}^{k} \lambda_s \mu^s \circ a^s \right\rangle = \sum_{s=1}^{k} \sum_{i=1}^{n} \langle \alpha_i a_i^*, \lambda_s \mu_i^s a_i^s\rangle = \sum_{s=1}^{k} \lambda_s \varrho_s \sum_{i=1}^{n} \alpha_i \mu_i^s = 0.$$

This equation holds if and only if $(\alpha_1, \ldots, \alpha_n) = \sigma(\omega_1, \ldots, \omega_n)$ with $\sigma \in \mathbb{R}$ and $\omega = (\omega_1, \ldots, \omega_n)$ any co-cycle of G. This assertion is a consequence of the fact that cycle basis and co-cycle basis are orthogonal complements in $\mathbb{R}^n$. Now it follows $x^* = (\sigma\omega_1 a_1^*, \ldots, \sigma\omega_n a_n^*) = \sigma a^* \circ \omega$ for any co-cycle ω. Consequently, as a result we get that statement (ii) holds if and only if $x^* \in L_{a,a*}$ thus $[\Phi_{a*}]^* = L_{a,a*}$. Analogously we get

$$[\Theta_a]^* = \{x \in \Phi_{a*} : \langle z^*, x\rangle = 0 \quad \text{for all } z^* \in \Theta_a\} = L_{a*,a}.$$

We observe that all considerations made in Sections 1.4 and 1.5 concerning convex functions and their conjugates remain also in case f_i are real valued convex functionals defined on Banach spaces E_i. We only consider closed convex functionals according to Definition 1.4.3. Similar to Lemma 1.5.2 and according to a theorem by Fenchel and Moreau [12], [28], [29] we have

$$f_i^{**}(x_i) = \sup_C \{\langle x_i^*, x_i\rangle - f_i^*(x_i^*)\} = f(x_i). \tag{3.6.16}$$

For it we also refer to [18].

Let $f_1, \ldots, f_n$ be closed convex functionals defined on $C_i \subseteq E_i$, $i = 1, \ldots, n$. We denote $C = C_1 \times \cdots \times C_n$ and

$$f(x) = \sum_{i=1}^{n} f_i(x_i).$$

Then it follows

$$\sup_C \{\langle x^*, x\rangle - f(x)\} = \sum_{i=1}^{n} \sup_{C_i} \{\langle x_i^*, x_i\rangle - f_i(x_i)\} = \sum_{i=1}^{n} f_i^*(x_i^*) = f^*(x^*).$$

The function f^* is defined on $C^* = C_1^* \times \cdots \times C_n^*$. Now we consider a Lagrangian form

$$\psi(x, x^*) = \langle x^*, x\rangle - f(x)$$

defined on $C \times \Theta_a$ and calculate:

$$\sup_C \psi(x, x^*) = \sup_C \{\langle x^*, x\rangle - f(x)\} = f^*(x^*),\ x^* \in C^* \cap \Theta_a,$$

$$\inf_{\Theta_a} \psi(x, x^*) = -f(x) + \inf_{\Theta_a} \langle x^*, x\rangle = \begin{cases} -f(x), & x \in C \cap L_{a*,a}, \\ -\infty, & x \notin C \cap L_{a*,a}. \end{cases}$$

Consequently for all $(x, x^*) \in (C \cap L_{a^*,a}) \times (C^* \cap \Theta_a)$ it holds $-f(x) \leqq f^*(x^*)$. The dual pair of problems we get in form of:

$$\text{Find } W = \sup\{-f(x): x \in C \cap L_{a^*,a}\}, \quad Z = C \cap L_{a^*,a}, \tag{P}$$

$$\text{find } W^* = \inf\{f^*(x^*): x^* \in C^* \cap \Theta_a\}, \quad Z^* = C^* \cap \Theta_a. \tag{D}$$

We call (P) and (D) *generalized transportation and potential problem* respectively.

The principle of duality we realize here by considering the transformation

$$f \to f^*, \quad C \to C^*, \quad \Theta_a \to [\Theta_a]^* = L_{a^*,a}. \tag{3.6.17}$$

According to (3.6.16) we get by application of this transformation twice

$$[f, C, \Theta_a]^{**} = [f^*, C^*, L_{a^*,a}]^* = [f^{**}, C^{**}, [L_{a^*,a}]^*] = [f, C, \Theta_a].$$

In case of generalized transportation and potential problems we have to modify our definition for subconsistence and subvalues.

Definition 3.6.3. (i) A sequence $\{x^k\} \subseteqq C$ is said to be *admissible* if there exist sequences $\{\bar{x}^k\} \in L_{a^*,a}$, $\{\bar{\bar{x}}^k\} \subseteqq E$ such that $x^k = \bar{x}^k + \bar{\bar{x}}^k$ and $\lim_{k\to\infty} \bar{\bar{x}}^k = 0$.

If the set $\hat{Z}$ of all admissible sequences is not empty then we call problem (P) *subconsistent* and define by

$$\omega = \sup_{\hat{Z}} \limsup_{k\to\infty} \{-f(x^k)\}$$

the *subvalue* of the transportation problem:

(ii) A sequence $\{x^{*k}\} \subseteqq C^*$ is said to be *admissible* if there exist sequences $\{\bar{x}^{*k}\} \subseteqq \Theta_a$, $\{\bar{\bar{x}}^{*k}\} \subseteqq E^*$ such that $x^{*k} = \bar{x}^{*k} + \bar{\bar{x}}^{*k}$ and $\lim_{k\to\infty} \bar{\bar{x}}^{*k} = 0$. If the set $\hat{Z}^*$ of admissible sequences is not empty then we call problem (D) *subconsistent* and define by

$$\omega^* = \inf_{\hat{Z}^*} \liminf_{k\to\infty} f^*(x^{*k})$$

the *subvalue* of the potential problem.

According to Definition 3.6.3 it is obvious that $W \leqq \omega$ and $\omega^* \leqq W^*$. There arises the question under which condition we get $W = \omega$ and $W^* = \omega^*$. Now we will discuss this problem.

Let z^* be any element contained in Θ_a. We consider the following problems:

$$\begin{aligned} &\text{Find } W^*(z^*) = \inf\{f^*(x^*): x^* \in Z^*(z^*)\}, \\ &\omega^*(z^*) = \inf \liminf_{k\to\infty} \left\{f^*(x^{*k}): \{x^{*k}\} \in \hat{Z}^*(z^*)\right\} \end{aligned} \tag{D_{z^*}}$$

with

$$Z^*(z^*) = \{x^*: x^* \in C^*, x^* - z^* \in \Theta_a\},$$

$$\hat{Z}^*(z^*) = \left\{\{x^{*k}\}: x^{*k} = \bar{x}^{*k} + \bar{\bar{x}}^{*k} \text{ and } \bar{x}^{*k} - z^* \in \Theta_a, \lim_{k\to\infty} \bar{\bar{x}}^{*k} = 0\right\}.$$

The real valued function $W^*(z^*)$ is obviously defined on $C^* - \Theta_a$. In case the problems (D_{z^*}) are not consistent we put $W^*(z^*) = \infty$ and $\omega^*(z^*) = \infty$. We define the closure of $W^*(z^*)$ by

$$\overline{W}^*(z^*) = \inf \liminf_{k\to\infty} \left\{W^*(z^{*k}): z^{*k} \in C^* - \Theta_a \text{ and } \lim_{k\to\infty} z^{*k} = z^*\right\}.$$

Now we prove

$$\overline{W}^*(z^*) = \omega^*(z^*). \tag{3.6.18}$$

Namely from $z^{*k} \in C^* - \Theta_a$ follows the existence of a sequence $x^{*k} \in C^*$ with $x^{*k} - z^{*k} \in \Theta_a$. We choose $z^{*k} \in C^* - \Theta_a$ such that $\lim_{k\to\infty} z^{*k} = z^*$ and consider

$$x^{*k} = (x^{*k} - z^{*k} + z^*) + (z^{*k} - z^*) = \bar{x}^{*k} + \bar{\bar{x}}^{*k}.$$

Then it follows $\lim_{k\to\infty} \bar{\bar{x}}^{*k} = 0$ and $\bar{x}^{*k} - z^* \in \Theta_a$ thus $\{x^{*k}\} \in \hat{Z}^*(z^*)$. Consequently we get

$$\begin{aligned}\omega^*(z^*) &= \inf \liminf_{k\to\infty} \{f^*(x^{*k}) : \{x^{*k}\} \in \hat{Z}^*(z^*)\}\\ &= \inf_{\{z^{*k}\}} \liminf_{k\to\infty} W^*(z^{*k}) = \overline{W}^*(z^*).\end{aligned} \tag{3.6.19}$$

Applying transformation (3.6.17) on (3.6.19) we get

$$\overline{W}(z) = \omega(z) \tag{3.6.20}$$

with

$$W(z) = \sup\{-f(x) : x \in Z(z)\};\ Z(z) = \{x : x \in C,\ x - z \in L_{a^*,a}\}$$

$$\omega(z) = \sup \limsup_{k\to\infty} \{-f(x^k) : \{x^k\} \in \hat{Z}(z)\};$$

$$\hat{Z}(z) = \left\{\{x^k\} : x^k = \bar{x}^k + \bar{\bar{x}}^k \text{ and } \bar{x}^k - z \in L_{a^*,a},\ \lim_{k\to\infty} \bar{\bar{x}}^k = 0\right\}$$

$$\overline{W}(z) = \sup_{\{z^k\}} \limsup_{k\to\infty} W(z^k),\ z^k \in C - L_{a^*,a} \text{ and } \lim_{k\to\infty} z^k = z.$$

The domain of definition of the function $W(z)$ is $C - L_{a^*,a}$. Since f^* is convex the function $\omega^*(z^*)$ is convex too and according (3.6.18) the function $\overline{W}^*(z^*)$ will be convex. Following a theorem by Ioffe and Tikhomirov [18] the convexity of $\overline{W}^*(z^*)$ guarantees with respect to (3.6.18)

$$\omega = \omega^*(0) = \overline{W}^*(0) = \sup_Z \{-f(x)\} = W.$$

Applying transformation (3.6.17) it follows immediately $\omega = W^*$. To summarize the results we now formulate the theorem:

Theorem 3.6.7. *If* $f(x) = \sum_{i=1}^{n} f_i(x_i)$ *is a closed convex function it holds:*

(i) *from the consistence of* (P) *it follows the subconsistence of* (D) *and* $W = \omega^* = \overline{W}^*(0)$;
(ii) *from the consistence of* (D) *it follows the subconsistence of* (P) *and* $W^* = \omega = \overline{W}(0)$.

Corrolary 3.6.3. *It is* $W = \omega$ *if and only if*

$$\sup_C \inf_{\Theta_a} \psi(x, x^*) = \inf_{\Theta_a} \sup_C \psi(x, x^*); \tag{3.6.21}$$

it is $W^* = \omega^*$ *if and only if*

$$\sup_{C^*} \inf_{L_{a^*,a}} \psi^*(x^*, x) = \inf_{L_{a^*,a}} \sup_{C^*} \psi^*(x^*, x) \tag{3.6.22}$$

with

$$\psi^*(x^*, x) = \langle x^*, x\rangle - f^*(x^*).$$

Proof: We get (3.6.22) immediately from (3.6.21) by application of transformation (3.6.17). Furthermore it is

$$\inf_{\Theta_a} \psi(x, x^*) = \begin{cases} -f(x); & x \in C \cap L_{a^*,a}, \\ -\infty; & x \notin C \cap L_{a^*,a}. \end{cases}$$

Consequently

$$W = \sup_Z \{-f(x)\} = \sup \{-f(x): x \in C\} = \sup_C \inf_{\Theta_a} \psi(x, x^*).$$

On the other hand it is

$$\sup_C \psi(x, x^*) = \begin{cases} f^*(x^*); & x^* \in C^* \cap \Theta_a, \\ \infty; & x^* \notin C^* \in \Theta_a, \end{cases}$$

thus

$$W^* = \inf_{Z^*} \{f^*(x^*)\} = \inf \{f^*(x^*): x^* \in \Theta_a\} = \inf_{\Theta_a} \sup_C \psi(x, x^*).$$

According to Theorem 3.6.7 it is $W^* = \omega$ hence $\omega = W$ if and only if (3.6.21) holds.

It is obvious that from $W = \omega$ and $W^* = \omega^*$ it follows $W = W^*$. Consequently the following statements are equivalent:

(i) $\sup_C \inf_{\Theta_a} \psi(x, x^*) = \inf_{\Theta_a} \sup_C \psi(x, x^*)$

(ii) $W = \omega$

(iii) $W = W^* = \overline{W}(0)$ (S)

(iv) $\overline{W}^*(0) = \omega^* = W^*$

(v) $\sup_{C^*} \inf_{L_{a^*,a}} \psi^*(x^*, x) = \inf_{L_{a^*,a}} \sup_{C^*} \psi^*(x^*, x)$.

At last we will present relations between saddle-point theorems, duality theorems and realizability of transportation and potential problems. For that we define a pair of sequences $(x^k, x^{*k}) \in C \times \Theta_a$ a *generalized saddle point* of $\psi(x, x^*)$ if

$$\limsup_{k\to\infty} \sup_C \psi(x, x^{*k}) \leqq \liminf_{k\to\infty} \inf_{\Theta_a} \psi(x^k, x^*)$$

for any $(x, x^*) \in C \times \Theta_a$. In case there exists a saddle point of $\psi(x, x^*)$ it follows obviously the existence of a generalized saddle point.

Theorem 3.6.8. *There exists a generalized saddle point of $\psi(x, x^*)$ if and only if the statements* (S) *holds.*

Proof: We assume one of the statements (S) to be true thus

$$\sup_C \inf_{\Theta_a} \psi(x, x^*) = \inf_{\Theta_a} \sup_C \psi(x, x^*).$$

Then there exist sequences $x^k \in C \cap L_{a^*,a}$, $x^{*k} \in C^* \cap \Theta_a$ such that

$$W = \lim_{k\to\infty} \{-f(x^k)\} = W^* = \lim_{k\to\infty} f^*(x^{*k}).$$

Consequently

$$\lim_{k\to\infty} \inf_{\Theta_a} \psi(x^k, x^*) = \lim_{k\to\infty} \sup_C \psi(x, x^{*k})$$

thus (x^k, x^{*k}) is a generalized saddle point.

To prove sufficiently, we assume

$$\limsup_{k\to\infty} \sup_C \psi(x, x^{*k}) \leqq \liminf_{k\to\infty} \inf_{\Theta_a} \psi(x^k, x^*). \tag{3.6.23}$$

We consider three cases: (i) $x^{*k} \notin C^*$. Then we get

$$\sup_C \psi(x, x^{*k}) = \sup_C \{\langle x^{*k}, x\rangle - f(x)\} = \infty$$

and with respect to (3.6.23) it follows

$$\liminf_{k\to\infty} \inf_{\Theta_a} \psi(x^k, x^*) = \infty.$$

But this is possible only for $x^k \in L_{a^*,a}$ thus

$$\liminf_{k\to\infty} \inf_{\Theta_a} \psi(x^k, x^*) = \liminf_{k\to\infty} \{-f(x^k)\} = W = \infty.$$

Furthermore from $W \leqq W^*$ it follows $W = W^* = \infty$.

(ii) Analogously we get $W = W^* = -\infty$ in case $x^k \notin L_{a^*,a}$.

(iii) We now assume $x^{*k} \in C^* \cap \Theta_a$. Then we get immediately $x^k \in C \cap L_{a^*,a}$ and by (3.6.23)

$$\begin{aligned} W^* &\leqq \limsup_{k\to\infty} f^*(x^{*k}) = \limsup_{k\to\infty} \sup_C \psi(x, x^{*k}) \\ &\leqq \liminf_{k\to\infty} \inf_{\Theta_a} \psi(x^k, x^*) = \liminf_{k\to\infty} \{-f(x^k)\} \leqq W. \end{aligned}$$

Since always $W \leqq W^*$ we finally get $W = W^*$ and this is equivalent to the statements (S). □

The existence of a saddle point in form of

$$\psi(x, \bar{x}^*) \leqq \psi(\bar{x}, \bar{x}^*) \leqq \psi(\bar{x}, x^*)$$

for all $(x, x^*) \in C \times \Theta_a$ implies the realizability of problems (P) and (D). Namely from the saddle-point inequality we get

$$\psi(\bar{x}, \bar{x}^*) = \max_C \{\langle \bar{x}^*, x\rangle - f(x)\} = \begin{cases} f^*(\bar{x}^*), & \bar{x}^* \in C^* \cap \Theta_a, \\ \infty, & \bar{x}^* \notin C^* \cap \Theta_a, \end{cases}$$

$$\psi(\bar{x}, \bar{x}^*) = \min_{\Theta_a} \{\langle x^*, \bar{x}\rangle - f(\bar{x})\} = \begin{cases} -f(\bar{x}), & \bar{x} \in C \cap L_{a^*,a}, \\ -\infty, & \bar{x} \notin C \cap L_{a^*,a} \end{cases}$$

thus

$$\psi(\bar{x}, \bar{x}^*) = -f(\bar{x}) \leqq W \leqq W^* \leqq f^*(\bar{x}^*) = \psi(\bar{x}, \bar{x}^*).$$

Consequently if (P) and (D) are realizable and $W = W^*$ it follows

$$f^*(\bar{x}^*) = W^* = \psi(\bar{x}, \bar{x}^*) = \sup_C \psi(x, \bar{x}^*) \geqq \psi(x, \bar{x}^*),$$

$$-f(\bar{x}) = W = \inf_{\Theta_a} \psi(\bar{x}, x^*) \leqq \psi(\bar{x}, x^*).$$

Hence $(\bar{x}, \bar{x}^*)$ is a saddle point of $\psi(x, x^*)$.

Now we will show with a simple example the existence of a duality-gap. We consider a graph with vertices $X = \{a, b, c\}$ and arcs $U = \{1, 2, 3\}$ such that $1 = (c, a)$, $2 = (c, b)$, $3 = (b, a)$. Obviously we get a cycle basis of $G = (X, U)$ in form of $\mu = (-1, 1, 1)$ and a co-cycle basis $\omega^1 = (1, 1, 0)$, $\omega^2 = (-1, 0, -1)$. Furthermore we choose $F = E^*$

$= \mathbb{R}^3$, $a^* = (1, 1, 1)$, $a = (1, 1, 1)$. Obviously it is $N_a = N_{a_1} \times N_{a_2} \times N_{a_3} = \{0\}$. Now we get

$$L_{a^*,a} = \{x \in \mathbb{R}^3 \colon x = (-\lambda, \lambda, \lambda),\ \lambda \in \mathbb{R}\}$$

$$\Theta_a = \{x^* \in \mathbb{R}^3 \colon x^* = (\alpha - \beta, \alpha, -\beta),\ \alpha, \beta \in \mathbb{R}\}.$$

Let

$$f_1(x_1) = -x_1, \qquad C_1 = [-1, \infty),$$

$$f_2(x_2) = -x_2, \qquad C_2 = (1, \infty),$$

$$f_3(x_3) = 2x_3, \qquad C_3 = (1, \infty).$$

The conjugate functions will take the form

$$f_1^*(x_1^*) = -x_1^* - 1, \qquad C_1^* = (-\infty, -1],$$

$$f_2^*(x_2^*) = x_2^* + 1, \qquad C_2^* = (-\infty, -1],$$

$$f_3^*(x_3^*) = x_3^* - 2, \qquad C_3^* = (-\infty, 2].$$

Now it follows

$$L_{a^*,a} \cap C = \{(-\lambda, \lambda, \lambda) \colon \lambda \leqq 1 \text{ and } \lambda > 1\} = \emptyset$$

thus (P) is not consistent hence $W = -\infty$. Furthermore we get

$$C^* \cap \Theta_a = \{(\alpha - \beta, \alpha, -\beta) \colon \alpha - \beta \leqq -1,\ \alpha \leqq -1,\ \beta \geqq -2\}$$

thus (D) is consistent. By Theorem 3.6.7 it follows the subconsistence of (P) and $W^* = \omega$. Truly it is

$$W^* = \inf_{C^* \cap \Theta_a} \{\beta - \alpha - 1 + \alpha + 1 - \beta - 2\} = -2.$$

An admissible sequence of problem (P) we can represent in form of

$$x^k = \bar{x}^k + \bar{\bar{x}}^k = (-1, 1, 1) + (0, 1/k, 1/k).$$

This sequence is optimal because

$$\omega = \sup_{\hat{z}} \limsup_{k \to \infty} \{-f(x^k)\} = \lim_{k \to \infty} \{-1 + 1 + 1/k - 2 - 2/k\} = -2.$$

3.7. Other problems of nonlinear programming

Generally if we speak about nonlinear programming, we mean to find an extremal value of a function over a given set of values of the variables. If the function is linear and the set can be represented by a linear algebraic system of equations and inequalities, then we have the case of linear programming.

In this section, we will make some remarks about a general nonlinear programming problem. Let G be a mapping from $\mathbb{R}^n$ into $\mathbb{R}^m$. Then G is a vector function which we can represent in the form

$$G(x) = \big(g_1(x), \ldots, g_m(x)\big).$$

Furthermore, we consider a function $f(x)$ defined on $K \subseteqq \mathbb{R}^n$ and put

$$\varphi(x, y) = \langle y, G(x)\rangle \quad \text{defined on } \mathbb{R}^n \times \mathbb{R}^m,$$

$$\alpha(x) = -f(x) \quad \text{defined on } K,$$

$$\beta(y) = 0 \quad \text{defined on } \mathbb{R}^{m+}.$$

Then the Lagrangian form will be

$$\psi(x, y) = \langle y, G(x)\rangle + f(x)$$

defined on $K \times L = K \times \mathbb{R}^{m+}$.

From the point of view of a general theory, we construct a dual pair of problems as follows. First, we find

$$\mu(x) = \inf_{\mathbb{R}^{m+}} \psi(x, y) = f(x) + \inf_{\mathbb{R}^{m+}} \langle y, G(x)\rangle.$$

Obviously,

$$\mu(x) = \begin{cases} f(x); \; x \in M = \{x \in K: G(x) \in \mathbb{R}^{m+}\} \\ -\infty; \; x \notin M. \end{cases}$$

and the domain of definition for μ is $K \subseteq \mathbb{R}^n$. Next, we find

$$\nu(y) = \sup_K \psi(x, y) = \sup_K \{\langle y, G(x)\rangle + f(x)\}$$

$$\nu(y) = \begin{cases} \sup_K \psi(x, y); \; y \in N = \left\{y \in \mathbb{R}^{m+}: \sup_K \psi(x, y) < \infty\right\} \\ +\infty; \; y \notin N \end{cases}$$

and ν is defined on $\mathbb{R}^{m+}$. Then the dual pair of problems take the form:

$$\text{Find } \tilde{W} = \sup_K \mu(x) = \sup_K \inf_{\mathbb{R}^{m+}} \psi(x, y), \tag{$\tilde{\mathrm{P}}$}$$

$$\text{find } \tilde{W}^* = \inf_{\mathbb{R}^{m+}} \nu(y) = \inf_{\mathbb{R}^{m+}} \sup_K \psi(x, y). \tag{$\tilde{\mathrm{D}}$}$$

Now, by definitions, given in Section 2.3,

$$\alpha^*(y) = \sup_K \{\langle y, G(x)\rangle + f(x)\} = \nu(y) = \alpha^*(y) - \beta(y)$$

$$\beta^*(x) = \inf_{\mathbb{R}^{m+}} \langle y, G(x)\rangle = \begin{cases} 0, & x \in M, \\ -\infty, & x \notin M \end{cases}$$

and therefore $\mu(x) = \beta^*(x) - \alpha(x)$.

If M and N are not empty, we call $x \in M$ and $y \in N$ *admissible* points. Then it is appropriate for us to consider instead of problems ($\tilde{\mathrm{P}}$) and ($\tilde{\mathrm{D}}$) corresponding problems (P) and (D) concerning only admissible points, where (P) and (D) have the following form:

$$\text{Find } W = \sup_M f(x), \tag{P}$$

$$\text{find } W^* = \inf_N \nu(y). \tag{D}$$

In general, these problems are not realizable and we cannot replace supremum by maximum and infium by minimum. This consideration leads us to a definition of an admissible sequence as we can find it in the work of GOLDSTEIN [14], [15].

Definition 3.7.1. We call a sequence

$$\mathscr{X} = \{x^k\} \subseteq M$$

an *admissible sequence* of problem (P) if

$$\lim_{k\to\infty} f(x^k)$$

exists inclusively of the case

$$\lim_{k\to\infty} f(x^k) = \infty .$$

We may now introduce the notation $f(\mathscr{X}) = \lim_{k\to\infty} f(x^k)$.

Definition 3.7.2. We call an admissible sequence $\mathscr{X}^0$ of problem (P) *optimal*, if

$$f(\mathscr{X}^0) = W = \sup_M f(x).$$

Similarly, we may define admissible sequences and optimally admissible sequences for the problems ($\tilde{\mathrm{P}}$), ($\tilde{\mathrm{D}}$) and (D). As a matter of fact, the problems (P), (D) and ($\tilde{\mathrm{P}}$), ($\tilde{\mathrm{D}}$) are not realizable in general, but there always exists an optimally admissible sequence. For example, the problem

$$\sup_M f(x) = \sup_{[1,\infty)} \left\{-\frac{1}{x}\right\} = 0$$

is obviously not realizable. But the sequence $\mathscr{X}^0 = \{n : n = 1, 2, \ldots\}$ is an optimally admissible sequence, because

$$f(\mathscr{X}^0) = \lim_{n\to\infty} \left\{-\frac{1}{n}\right\} = \sup_{[1,\infty)} \left\{-\frac{1}{x}\right\} = 0 .$$

Now, we have the following theorem concerning optimally admissible sequences which we shall refer to as optimal sequences later.

Theorem 3.7.1. *Problems* (P), (D) *and* ($\tilde{\mathrm{P}}$), ($\tilde{\mathrm{D}}$) *are equivalent in the following sense.*

(i) $W = \tilde{W}$ *and* $W^* = \tilde{W}^*$.

(ii) *If* $\mathscr{X}$ *and* $\mathscr{Y}$ *are optimal sequence of* (P) *and* (D) *respectively, then* $\mathscr{X}$ *and* $\mathscr{Y}$ *are also optimal sequences of* ($\tilde{\mathrm{P}}$) *and* ($\tilde{\mathrm{D}}$).

(iii) *If* $\mathscr{X}$ *and* $\mathscr{Y}$ *are optimal sequences of* ($\tilde{\mathrm{P}}$) *and* ($\tilde{\mathrm{D}}$) *respectively with*

$$\tilde{W} = \mu(\mathscr{X}) = \lim_{k\to\infty} \mu(x^k) > -\infty,$$

$$\tilde{W}^* = \nu(\mathscr{Y}) = \lim_{k\to\infty} \nu(y^k) < +\infty$$

then $\mathscr{X}$ *and* $\mathscr{Y}$ *are also optimal sequences of* (P) *and* (D).

Proof: (i) Assume that $W = -\infty$. It follows necessarily that $M = \emptyset$. Consequently, $\mu(x) = -\infty$ and thus $\tilde{W} = -\infty$. In case $W > -\infty$, it follows that $M \neq \emptyset$. Then we get

$$\tilde{W} = \sup_K \mu(x) = \sup_M f(x) = W .$$

Similarly we can show that $W^* = \tilde{W}^*$.

(ii) Let $\mathscr{X}$ and $\mathscr{Y}$ be optimal sequences of (P) and (D) respectively. According to the definition, the sets M and N are not empty and we have

$$W = \sup_M f(x) = f(\mathscr{X}) = \lim_{k\to\infty} f(x^k) > -\infty,$$

$$W^* = \inf_N \nu(y) = \nu(\mathscr{Y}) = \lim_{k\to\infty} \nu(y^k) < \infty$$

and thus

$$W = \lim_{k\to\infty} f(x^k) = \lim_{k\to\infty} \mu(x^k) = \mu(\mathcal{X}),$$

$$W^* = \lim_{k\to\infty} \nu(y^k) = \nu(\mathcal{Y}).$$

Using statement (i) we get

$$\mu(\mathcal{X}) = W, \qquad \nu(\mathcal{Y}) = W^*.$$

This means $\mathcal{X}$ and $\mathcal{Y}$ are optimal sequences of $(\tilde{\mathrm{P}})$ and $(\tilde{\mathrm{D}})$.

(iii) If $\mathcal{X}$ and $\mathcal{Y}$ are optimal sequences of $(\tilde{\mathrm{P}})$ and $(\tilde{\mathrm{D}})$ respectively, then, by assumption,

$$\tilde{W} = \mu(\mathcal{X}) = \lim_{k\to\infty} \mu(x^k) > -\infty,$$

$$\tilde{W}^* = \nu(\mathcal{Y}) = \lim_{k\to\infty} \nu(y^k) < \infty.$$

Thus the sets M and N are nonempty and $x^k \in M$ and $y^k \in N$. According to statement (i) of the theorem it follows that

$$\tilde{W} = \mu(\mathcal{X}) = \lim_{k\to\infty} \mu(x^k) = \lim_{k\to\infty} f(x^k) = f(\mathcal{X}) = W,$$

$$\tilde{W}^* = \nu(\mathcal{Y}) = \lim_{k\to\infty} \nu(y^k) = W^* = \inf_N \nu(y).$$

Thus

$$W = \sup_M f(x) = f(\mathcal{X})$$

$$W^* = \inf_N \nu(y) = \nu(\mathcal{Y}).$$

This means $\mathcal{X}$ and $\mathcal{Y}$ are optimal sequences of (P) and (D). Analogous to Theorem 2.3.1, we have the following theorem.

Theorem 3.7.2. *The following three statements are equivalent:*

(i) *There exist points* $(x^0, y^0) \in K \times \mathbb{R}^{m+}$ *such that for all* $(x, y) \in K \times \mathbb{R}^{m+}$,

$$\langle y^0, G(x)\rangle + f(x) \leqq \langle y^0, G(x^0)\rangle + f(x^0) \leqq \langle y, G(x^0)\rangle + f(x^0).$$

(ii) *The sets M and N are not empty, the values W and W^* are finite and the problems* (P) *and* (D) *are realizable and*

$$W = \max_M f(x) = \min_N \nu(y) = W^*.$$

(iii) *There exist points* $(x^0, y^0) \in M \times N$ *with*

$$0 = \langle y^0, G(x^0)\rangle = \nu(y^0) - f(x^0).$$

According to Theorem 3.7.1, we can substitute statement (ii) of Theorem 3.7.2 by a corresponding duality statement with respect to the problems $(\tilde{\mathrm{P}})$ and $(\tilde{\mathrm{D}})$. The relation $\tilde{W} = \tilde{W}^*$ is obviously equivalent to

$$\inf_{\mathbb{R}^{m+}} \sup_K \varphi(x, y) = \sup_K \inf_{\mathbb{R}^{m+}} \varphi(x, y).$$

Furthermore, statement (iii) of Theorem 3.7.2 implies that

$$\widetilde{W} = \max_K \mu(x) = \max_M f(x) = f(x^0) = \nu(y^0) = \widetilde{W}^* = \min_{\mathbb{R}^{m+}} \nu(y).$$

Consequently,

$$\min_{\mathbb{R}^{m+}} \sup_K \psi(x, y) = \max_K \inf_{\mathbb{R}^{m+}} \psi(x, y).$$

Now, we assume that the functions $f(x)$, $g_1(x)$, ..., $g_m(x)$ possess first continuous partial derivatives. Then, by Theorem 1.4.7, all these functions are Fréchet differentiable and their gradients are Fréchet derivatives. Now, we put $K = \mathbb{R}^{n+}$ and assume that for one point $x^1 \in \mathbb{R}^{n+}$ we have

$$\alpha^*(y) = \nu(y) = \max_{\mathbb{R}^{n+}} \{\langle y, G(x)\rangle + f(x)\} = \langle y, G(x^1)\rangle + f(x^1).$$

Then, according to Theorem 1.4.8, we have necessarily

$$\text{grad}_x \left(\sum_{i=1}^m \eta_i g_i(x^1) + f(x^1)\right) \leqq 0,$$

$$\left\langle \text{grad}_x \left(\sum_{i=1}^m \eta_i g_i(x^1) + f(x^1)\right), x^1 \right\rangle = 0.$$

Consequently, the domain of definition for the function $\nu(y) = \langle y, G(x^1)\rangle + f(x^1)$ is the set

$$N = \left\{ y \in \mathbb{R}^{m+} : \sum_{i=1}^m \eta_i \frac{\partial g_i(x^1)}{\partial \xi_j} \leqq \frac{-\partial f(x^1)}{\partial \xi_j};\quad \sum_{j=1}^n \xi_j^1 \sum_{i=1}^m \eta_i \frac{\partial g_i}{\partial \xi_j} = -\sum_{j=1}^n \xi_j^1 \frac{\partial f}{\partial \xi_j} \right\}.$$

In this case, the dual pair of problems appears in the following form:

$$\text{Find } W = \max_M f(x), \tag{P}$$

$$\text{find } W^* = \min_N \left\{ \sum_{i=1}^m \eta_i g_i(x^1) + f(x^1) \right\}. \tag{D}$$

If (x^0, y^0) is a saddle point for $\psi(x, y)$ with respect to $\mathbb{R}^{n+} \times \mathbb{R}^{m+}$, it follows from Theorem 3.7.2 that

$$W = f(x^0) = \sum_{i=1}^n \eta_i^0 g_i(x^1) + f(x^1) = W^*$$

and this is equivalent to

$$\min_N \max_{\mathbb{R}^{n+}} \psi(x, y) = \max_M \min_{\mathbb{R}^{m+}} \psi(x, y).$$

Theorem 3.7.2 means that either for a dual pair of problems, all the statements (i), (ii), (iii) are true or none of those statements holds. If none of the statements is true, then it does not necessarily mean that the problems are not solvable.

Now consider the following example with

$$K = \mathbb{R},\quad L = \mathbb{R}^+,\quad f(x) = x,\quad G(x) = -x^2.$$

(i) Obviously, the Lagrangian form

$$\psi(x, y) = x - yx^2$$

does not possess a saddle point with respect to $R \times \mathbb{R}^+$.

(ii) We have

$$\mu(x) = \inf_{\mathbb{R}^+} \varphi(x, y) = x + \inf_{\mathbb{R}^+} \{-yx^2\} = \begin{cases} x & \text{for } x \in M = \{0\}, \\ -\infty & \text{for } x \neq 0, \end{cases}$$

$$\nu(y) = \sup_{\mathbb{R}} \varphi(x, y) = \sup_{\mathbb{R}} \{x - yx^2\} = \begin{cases} \frac{1}{4y};\ y \in N = \{y \in \mathbb{R}^+ : y \neq 0\}, \\ \infty;\ y \notin N. \end{cases}$$

The dual pair of problems appears now in the form

$$\widetilde{W} = \sup_{\mathbb{R}} \mu(x) = \sup_{M} \mu(x) = 0 = W, \tag{P}$$

$$\widetilde{W}^* = \inf_{\mathbb{R}^+} \nu(y) = \inf_{N} \frac{1}{4y} = 0 = W^*. \tag{D}$$

We notice that $W = W^*$, but the primal problem is realizable and $x^0 = 0$ is an optimal program while the dual problem is not realizable.

But $\mathscr{Y} = \{y_n = n\}$ is an optimal sequence, because

$$\lim_{n\to\infty} \frac{1}{4n} = 0 = \nu(\mathscr{Y}) = W^*.$$

(iii) Also the statement (iii) of Theorem 3.7.2 is not true in this special case, because for any element $y \in N$, we have

$$\nu(y) - f(x^0) = \nu(y) = \frac{1}{4y} \neq 0.$$

Moreover, this example also tells us that we can expect a generalized theorem of the form of Theorem 3.7.2 if we can find some suitable generalizations of the ideas of saddle point and solution.

Definition 3.7.3. We call a sequence $(\mathscr{X}^0, \mathscr{Y}^0)$ with $\mathscr{X}^0 = \{x_0^k\} \subseteq K$ and $\mathscr{Y}^0 = \{y_0^k\} \subseteq \mathbb{R}^{m+}$ *a generalized saddle point* of

$$\varphi(x, y) = \langle y, G(x)\rangle + f(x)$$

with respect to $K \times \mathbb{R}^{m+} \subseteq \mathbb{R}^n \times \mathbb{R}^m$, if we have for any $(x, y) \in K \times \mathbb{R}^{m+}$,

$$\lim_{k\to\infty} \sup \sup_{K} \varphi(x, y_0^k) \leq \lim_{k\to\infty} \inf \inf_{\mathbb{R}^{m+}} \varphi(x_0^k, y). \tag{3.7.1}$$

Lemma 3.7.1. *The sequence* $(\mathscr{X}^0, \mathscr{Y}^0)$ *is a generalized saddle point of* φ *if and only if*

$$\lim_{k\to\infty} \sup_{K} \varphi(x, y_0^k) = \lim_{k\to\infty} \varphi(x_0^k, y_0^k) = \lim_{k\to\infty} \inf_{\mathbb{R}^{m+}} \varphi(x_0^k, y). \tag{3.7.2}$$

Proof: Obviously,

$$\lim_{k\to\infty} \sup \varphi(x^k, y^k) \leq \lim_{k\to\infty} \sup \sup_{K} \varphi(x, y^k)$$

and

$$\lim_{k\to\infty} \inf \varphi(x^k, y^k) \geq \lim_{k\to\infty} \inf \inf_{\mathbb{R}^{m+}} \varphi(x^k, y).$$

From this, we get for a generalized saddle point $(\mathscr{X}^0, \mathscr{Y}^0)$

$$\limsup_{k\to\infty} \psi(x_0^k, y_0^k) \leqq \limsup_{k\to\infty} \sup_K \psi(x, y_0^k)$$
$$\leqq \liminf_{k\to\infty} \inf_{\mathbb{R}^{m+}} \psi(x_0^k, y) \leqq \liminf_{k\to\infty} \psi(x_0^k, y_0^k).$$

Because we always have

$$\liminf_{k\to\infty} \psi(x_0^k, y_0^k) \leqq \limsup_{k\to\infty} \psi(x_0^k, y_0^k)$$

it follows that

$$\limsup_{k\to\infty} \psi(x_0^k, y_0^k) = \liminf_{k\to\infty} \psi(x_0^k, y_0^k) = \lim_{k\to\infty} \psi(x_0^k, y_0^k)$$
$$= \limsup_{k\to\infty} \sup_K \psi(x, y_0^k) = \liminf_{k\to\infty} \inf_{\mathbb{R}^{m+}} (x_0^k, y). \tag{3.7.3}$$

Furthermore it is obvious that

$$\limsup_{k\to\infty} \inf_{\mathbb{R}^{m+}} \psi(x^k, y) \leqq \liminf_{k\to\infty} \sup_K \psi(x, y^k).$$

Thus,

$$\limsup_{k\to\infty} \sup_K \psi(x, y_0^k) \leqq \liminf_{k\to\infty} \inf_{\mathbb{R}^{m+}} \psi(x_0^k, y)$$
$$\leqq \limsup_{k\to\infty} \inf_{\mathbb{R}^{m+}} \psi(x_0^k, y) \leqq \liminf_{k\to\infty} \sup_K \psi(x, y_0^k)$$

and

$$\limsup_{k\to\infty} \inf_{\mathbb{R}^{m+}} \psi(x_0^k, y) \leqq \liminf_{k\to\infty} \sup_K \psi(x, y_0^k)$$
$$\leqq \limsup_{k\to\infty} \sup_K \psi(x, y_0^k) \leqq \liminf_{k\to\infty} \inf_{\mathbb{R}^{m+}} \psi(x_0^k, y).$$

From this follows

$$\limsup_{k\to\infty} \sup_K \psi(x, y_0^k) = \liminf_{k\to\infty} \sup_K \psi(x, y_0^k) = \lim_{k\to\infty} \sup_K \psi(x, y_0^k)$$

and

$$\limsup_{k\to\infty} \inf_{\mathbb{R}^{m+}} \psi(x_0^k, y) = \liminf_{k\to\infty} \inf_{\mathbb{R}^{m+}} \psi(x_0^k, y) = \lim_{k\to\infty} \inf_{\mathbb{R}^{m+}} \psi(x_0^k, y).$$

This result together with (3.7.3) implies (3.7.2) and consequently the relations (3.7.2) and (3.7.1) are equivalent. □

If a generalized saddle point is a stationary sequence $\{(x_0^k, y_0^k)\} = \{(x^0, y^0)\}$, $k = 1, 2, \ldots$, then it follows immediately from (3.7.2) that the ordinary saddle-point inequality holds.

Lemma 3.7.2. *The function $\psi(x, y)$ possesses a generalized saddle point if and only if for any $\varepsilon > 0$, there exist elements $(x^\varepsilon, y^\varepsilon) \in K \times \mathbb{R}^{m+}$ such that for any $(x, y) \in K \times \mathbb{R}^{m+}$, we have*

$$\psi(x, y^\varepsilon) - \varepsilon \leqq \psi(x^\varepsilon, y^\varepsilon) \leqq \psi(x^\varepsilon, y) + \varepsilon. \tag{3.7.4}$$

Proof: Suppose that the relation (3.7.4) holds. Then there exists for $n = 1, 2, \ldots$ a sequence $(x^n, y^n) \in K \times \mathbb{R}^{m+}$, with

$$\psi(x, y^n) - \frac{1}{n} \leqq \psi(x^n, y^n) \leqq \psi(x^n, y) + \frac{1}{n}.$$

This implies

$$\sup_K \psi(x, y^n) - \frac{1}{n} \leq \inf_{\mathbb{R}^{m+}} \psi(x^n, y) + \frac{1}{n}.$$

Thus

$$\limsup_{n\to\infty} \sup_K \psi(x, y^n) \leq \liminf_{n\to\infty} \inf_{\mathbb{R}^{m+}} \psi(x^n, y).$$

Consequently $(\mathcal{X}, \mathcal{Y}) = \{(x^n, y^n)\}$ is, according to the Definition 3.7.3, a generalized saddle point. If ψ possesses a generalized saddle point, then, by Lemma 3.7.1, corresponding to a generalized saddle point $(\mathcal{X}, \mathcal{Y})$, we have

$$\lim_{k\to\infty} \sup_K \psi(x, y^k) = \lim_{k\to\infty} \psi(x^k, y^k) = \lim_{k\to\infty} \inf_{\mathbb{R}^{m+}} \psi(x^k, y). \tag{3.7.5}$$

Now, we put

$$A = \lim_{k\to\infty} \sup_K \psi(x, y^k), \qquad B = \lim_{k\to\infty} \psi(x^k, y^k), \qquad C = \lim_{k\to\infty} \inf_{\mathbb{R}^{m+}} \psi(x^k, y).$$

Then, for any $\varepsilon > 0$, there exists a number $k(\varepsilon)$ such that for $k > k(\varepsilon)$, the following inequalities are true

$$\begin{aligned} A - \varepsilon/2 &\leq \sup_K \psi(x, y^k) \leq A + \varepsilon/2, \\ B - \varepsilon/2 &\leq \psi(x^k, y^k) \leq B + \varepsilon/2, \\ C - \varepsilon/2 &\leq \inf_{\mathbb{R}^{m+}} \psi(x^k, y) \leq C + \varepsilon/2. \end{aligned} \tag{3.7.6}$$

We choose any pair (x^{k_0}, y^{k_0}) with $k_0 > k(\varepsilon)$ and put $x^{k_0} = x^\varepsilon$, $y^{k_0} = y^\varepsilon$. Then we get from (3.7.5) and (3.7.6),

$$\psi(x^\varepsilon, y^\varepsilon) \leq B + \varepsilon/2 = C + \varepsilon/2 \leq \inf_{\mathbb{R}^{m+}} \psi(x^\varepsilon, y) + \varepsilon \leq \psi(x^\varepsilon, y) + \varepsilon$$

and

$$\psi(x^\varepsilon, y^\varepsilon) \geq B - \varepsilon/2 = A - \varepsilon/2 \geq \sup_K \psi(x, y^\varepsilon) - \varepsilon \geq \psi(x, y^\varepsilon) - \varepsilon.$$

But, these two inequalities together implies (3.7.4). □

Theorem 3.7.3. *The following two statements are equivalent:*

(i) *$\psi(x, y)$ possesses a generalized saddle point with respect to $K \times \mathbb{R}^{m+}$.*
(ii) $\widetilde{W} = \sup_K \mu(x) = \inf_{\mathbb{R}^{m+}} \nu(y) = \widetilde{W}^*$.

Proof: If ψ possesses a generalized saddle point (x^k, y^k), then, according to (3.7.2),

$$\widetilde{W}^* = \inf_{\mathbb{R}^{m+}} \nu(y) \leq \lim_{k\to\infty} \nu(y^k) = \lim_{k\to\infty} \mu(x^k) \leq \sup_K \mu(x) = \widetilde{W}.$$

Furthermore, we have, for any $(x, y) \in K \times \mathbb{R}^{m+}$

$$\mu(x) = \inf_{\mathbb{R}^{m+}} \psi(x, y) \leq \sup_K \psi(x, y) = \nu(y).$$

Consequently,

$$\widetilde{W} = \sup_K \mu(x) \leq \inf_{\mathbb{R}^{m+}} \nu(y) = \widetilde{W}^*.$$

Thus $\widetilde{W} = \widetilde{W}^*$.

Now we assume $\widetilde{W} = \widetilde{W}^*$ and let $\mathcal{X} = \{x^k\} \subseteqq K$, $\mathcal{Y} = \{y^k\} \subseteqq \mathbb{R}^{m+}$ be the optimal admissible sequences of problems $(\tilde{\mathrm{P}})$ and $(\tilde{\mathrm{D}})$. From this follows

$$\widetilde{W} = \lim_{k\to\infty} \mu(x^k) = \lim_{k\to\infty} \inf_{\mathbb{R}^{m+}} \varphi(x^k, y) = \widetilde{W}^* = \lim_{k\to\infty} \nu(y^k) = \lim_{k\to\infty} \sup_{K} \varphi(x, y^k).$$

This implies the inequality (3.7.1) and thus (x^k, y^k) is a generalized saddle point. □

Using these generalized concepts, we get the following generalization of Theorem 3.7.3.

Theorem 3.7.4. *If M and N are nonempty sets, then the following three statements are equivalent:*

(i) *$\varphi(x, y)$ possesses a generalized saddle point with respect to $K \times \mathbb{R}^{m+}$.*

(ii) $W = \sup_M f(x) = \inf_N \nu(y) = W^*$.

(iii) *There exist admissible sequences*

$$\mathcal{X} = \{x^k\} \subseteqq M, \qquad \mathcal{Y} = \{y^k\} \subseteqq N$$

such that

$$f(\mathcal{Y}) = \lim_{k\to\infty} f(x^k) = \lim_{k\to\infty} \nu(y^k) = \nu(\mathcal{Y}),$$

$$\lim_{k\to\infty} \langle y^k, G(x^k)\rangle = 0.$$

Proof: According to Theorem 3.7.1, we have

$$W = \sup_M f(x) = \sup_K \mu(x) = \widetilde{W},$$

$$W^* = \inf_N \nu(y) = \inf_{\mathbb{R}^{m+}} \nu(y) = \widetilde{W}^*.$$

Then it follows immediately from Theorem 3.7.3 that statements (i) and (ii) are equivalent. Now we assume that statement (i) is true and $(\mathcal{X} = \{x^k\}, \mathcal{Y} = \{y^k\})$ is a generalized saddle point. By Lemma 3.7.1, we have

$$\lim_{k\to\infty} \sup_K \varphi(x, y^k) = \lim_{k\to\infty} \nu(y^k) = \lim_{k\to\infty} \varphi(x^k, y^k) = \lim_{k\to\infty} \inf_{\mathbb{R}^{m+}} \varphi(x^k, y) = \lim_{k\to\infty} \mu(x^k).$$

Because M and N are not empty, we have $\mu(x^k) = f(x^k)$ and

$$\lim_{k\to\infty} \mu(x^k) = \lim_{k\to\infty} f(x^k) = \lim_{k\to\infty} \nu(y^k).$$

Furthermore, it follows from Lemma 3.7.2 that for any $(x, y) \in K \times \mathbb{R}^{m+}$,

$$f(x) + \langle y^k, G(x)\rangle - \frac{1}{k} \leqq f(x^k) + \langle y^k, G(x^k)\rangle \leqq f(x^k) + \langle y, G(x^k)\rangle + \frac{1}{k}.$$

Because $x^k \in M$, we have $G(x^k) \geqq 0$ and we get for $y = 0$,

$$0 \leqq \langle y^k, G(x^k)\rangle \leqq \frac{1}{k}.$$

Thus,

$$\lim_{k\to\infty} \langle y^k, G(x^k)\rangle = 0$$

which is statement (iii).

To prove the converse, we assume that $\mathcal{X} = \{x^k\} \subseteqq M$, $\mathcal{Y} = \{y^k\} \subseteqq N$ are admissible sequences with the property

$$\lim_{k\to\infty} f(x^k) = \lim_{k\to\infty} \nu(y^k),$$

$$\lim_{k\to\infty} \langle y^k, G(x^k)\rangle = 0.$$

From this follows

$$\lim_{k\to\infty} f(x^k) = \lim_{k\to\infty} \varphi(x^k, y^k)$$

and because $x^k \in M$,

$$\lim_{k\to\infty} f(x^k) = \lim_{k\to\infty} \mu(x^k).$$

Now we get,

$$\lim_{k\to\infty} f(x^k) = \lim_{k\to\infty} \mu(x^k) = \lim_{k\to\infty} \inf_{\mathbb{R}^{m+}} \varphi(x^k, y)$$

$$= \lim_{k\to\infty} \varphi(x^k, y^k) = \lim_{k\to\infty} \nu(y^k) = \lim_{k\to\infty} \sup_{K} \varphi(x, y^k)$$

which is the relation (3.7.2).

Thus $(\mathcal{X}, \mathcal{Y}) = \{(x^k, y^k)\}$ is a generalized saddle point of $\varphi(x, y)$.

We remark that the sets M and N can never be both empty in case $\widetilde{W} = \widetilde{W}^*$.

For example, if $M = \emptyset$, it follows that

$$\widetilde{W} = \sup_{K} \mu(x) = -\infty = \inf_{\mathbb{R}^{m+}} \nu(y) = \widetilde{W}^*.$$

But, $\widetilde{W}^* = -\infty$ implies $N \neq \emptyset$.

If we furthermore demand that the functions $f, g_1, \ldots, g_m$ are differentiable, we can find other statements equivalent to (i), (ii) and (iii) of Theorem 3.7.2. To do so, we need to make some preparations.

Definition 3.7.4. A real valued function $g(x)$ defined on $\mathbb{R}^n$ is called *linear-affine*, if $g(x) - g(0)$ is linear.

Lemma 3.7.3. *$g_1, g_2, \ldots, g_r$ are convex functions, defined on a convex set $C \subseteqq \mathbb{R}^n$. If the system $g_i(x) < 0$, $i = 1, 2, \ldots, r$, does not possess a solution in C, then there exists a function $g(x)$ expressible in the form*

$$g(x) = \sum_{k=1}^{n+1} \alpha_k g_{i_k}(x) \tag{3.7.7}$$

with $\alpha_k \geqq 0$ and $\sum_{k=1}^{n+1} \alpha_k = 1$, where $g(x)$ has the property $\inf_{C} g(x) \geqq 0$. In case $r \leqq n+1$, the sum (3.7.7) involves all the functions $g_1, \ldots, g_r$.

Proof: Let us consider the set

$$S = \{y = (\eta_1, \ldots, \eta_r) \in \mathbb{R}^r : \exists x \in C : g_i(x) < \eta_i,\ i = 1, 2, \ldots, r\}.$$

If $g_i(x) < 0$ possesses no solution, then $y = 0$ is not contained in S.

Furthermore, the set S is convex. Namely, if $y, y^1 \in S$ and $0 \leqq \lambda \leqq 1$, then there exist elements $x, x^1 \in C$ such that

$$g_i(x) < \eta_i, \; g_i(x^1) \leqq \eta_i^1, \; i = 1, \ldots, r,$$

where $\lambda x + (1 - \lambda) x^1 \in C$ and

$$g_i\big(\lambda x + (1 - \lambda) x^1\big) \leqq \lambda g_i(x) + (1 - \lambda) g_i(x^1) < \lambda \eta_i + (1 - \lambda) \eta_i^1$$

i.e.

$$\lambda y + (1 - \lambda) y^1 \in S.$$

According to the separation Theorem 1.2.3, there exists a vector $a \in \mathbb{R}^r$ with $a \neq 0$ such that

$$\langle a, y \rangle = \sum_{i=1}^{r} \alpha_i \eta_i \geqq 0$$

for any $y \in S$.

Also we notice that, for any $x \in C$ and any real positive numbers $\lambda_1, \ldots, \lambda_r$ the vector $y \in \mathbb{R}^r$ with components $\eta_i = g_i(x) + \lambda_i$ is contained in S. From this follows

$$\langle a, y \rangle = \sum_{i=1}^{r} \alpha_i g_i(x) + \sum_{i=1}^{r} \alpha_i \lambda_i \geqq 0$$

for any $x \in C$. Because this inequality is true for any real numbers $\lambda_1, \lambda_2, \ldots, \lambda_r$ with $\lambda_i > 0$ it follows necessarily that

$$\sum_{i=1}^{r} \alpha_i g_i(x) \geqq 0 \quad \text{for any } x \in C \tag{3.7.8}$$

and $a = (\alpha_1, \ldots, \alpha_r) \geqq 0$.

Without any loss of generality, we assume $\sum_{i=1}^{r} \alpha_i = 1$; otherwise, we divide α_i by $\sum_{i=1}^{r} \alpha_i > 0$.

In case $r \leqq n + 1$, we put

$$g(x) = \sum_{i=1}^{r} \alpha_i g_i(x)$$

and from (3.7.8) follows $\inf_C g(x) \geqq 0$. Next, we consider the case $r > n + 1$. Because the functions g_i's are convex, the sets

$$C_i = \{x \in C : g_i(x) < 0\}, \qquad i = 1, 2, \ldots, r,$$

are convex. Since $g_i(x) < 0$, $i = 1, 2, \ldots, r$, does not possess a solution, we have $\sum_{i=1}^{r} C_i = \emptyset$. By Theorem 1.1.8 (Helly), there exist indices $i_1, \ldots, i_{n+1}$ such that $\sum_{k=1}^{n+1} C_{i_k} = \emptyset$. Consequently, the system $g_{i_k}(x) < 0$, $k = 1, \ldots, n + 1$, is not solvable in C. According to the first part of the proof, there exist numbers $\alpha_1, \ldots, \alpha_{n+1} \geqq 0$ with $\sum_{k=1}^{n+1} \alpha_k = 1$ such that

$$\sum_{k=1}^{n+1} \alpha_i g_{i_k}(x) \geqq 0 \quad \text{for any } x \in C.$$

Again, we put

$$g(x) = \sum_{k=1}^{n+1} \alpha_i g_{i_k}(x)$$

and we get $\inf_C g(x) \geqq 0$ which proves the Lemma. □

Applying Lemma 3.7.3, we derive the well known theorem of FARKAS-MINKOWSKI [11].

Theorem 3.7.5 (Farkas-Minkowski). *Given $m + 1$ concave functions $f(x)$, $g_1(x), \ldots, g_m(x)$ in $\mathbb{R}^n$. Assume the existence of an index $j \leqq m$ such that the functions $g_i(x)$, $j < i \leqq m$, are linear-affine. If*

$$\begin{aligned} &g_i(x) \geqq 0, \quad i = 1, \ldots, m, \\ &f(x) > 0 \end{aligned} \tag{3.7.9}$$

is not solvable in $\mathbb{R}^n$, but

$$\begin{aligned} &g_i(x) > 0, \quad 1 \leqq i \leqq j, \\ &g_i(x) \geqq 0, \quad j < i \leqq m, \end{aligned} \tag{3.7.10}$$

possesses a solution $x \in \mathbb{R}^n$, then there exists a vector $y \in \mathbb{R}^{m+}$ with $y \neq 0$ such that

$$\psi(x, y) = f(x) + \langle y, G(x) \rangle = f(x) + \sum_{i=1}^{m} \eta_i g_i(x) \leqq 0$$

for any $x \in \mathbb{R}^n$.

Proof: According to Lemma 3.7.3, there exists a function

$$g(x) = \alpha_0 f(x) + \sum_{i=1}^{m} \alpha_i g_i(x)$$

with $\alpha_i \geqq 0$, $i = 0, 1, \ldots, m$ and $\sum_{i=0}^{m} \alpha_i = 1$ such that

$$g(x) = \alpha_0 f(x) + \sum_{i=1}^{m} \alpha_i g_i(x) \leqq 0 \tag{3.7.11}$$

for any $x \in \mathbb{R}^n$. In case $\alpha_0 \neq 0$, we divide the inequality by α_0 and put

$$y = (\eta_1, \ldots, \eta_m) = \left(\frac{\alpha_1}{\alpha_0}, \ldots, \frac{\alpha_m}{\alpha_0}\right) \in \mathbb{R}^{m+}.$$

Then it follows that

$$f(x) + \langle y, G(x) \rangle \leqq 0$$

for any $x \in \mathbb{R}^n$. Now, we will show that the case $\alpha_0 = 0$ is not possible. Namely if x^0 is a solution of (3.7.10), we get, in case $\alpha_0 = 0$

$$g(x^0) = \sum_{i=1}^{m} \alpha_i g_i(x^*) \geqq 0.$$

On the other hand according to (3.7.11)

$$g(x^0) = \sum_{i=1}^{m} \alpha_i g_i(x^*) \leqq 0.$$

Thus

$$g(x^0) = \sum_{i=1}^{m} \alpha_i g_i(x^0) = 0$$

and $\alpha_i = 0$ for $1 \leqq i \leqq j$.

Now, we get

$$g(x) = \sum_{i=j+1}^{m} \alpha_i g_i(x).$$

According to the assumption of the theorem, the function $g(x)$ must be linear-affine. But in this case, $g(x)$ must be zero for any $x \in \mathbb{R}^n$, because only in this case, a linear-affine function can satisfy the condition, $g(x^0) = 0$, $g(x) \leqq 0$, $x \in \mathbb{R}^n$.

We will now see that the fact that $g(x) = 0$ for any $x \in \mathbb{R}^n$ leads to contradiction. Using Corollary 1.2.2 of Theorem 1.2.8, we consider $m - j$ closed convex sets

$$C_i = \{x \in \mathbb{R}^n : g_i(x) \geqq 0\}, \quad i = j + 1, \ldots, m,$$

and the convex set

$$C = \{x \in \mathbb{R}^n : f(x) > 0,\ g_i(x) \geqq 0,\ i = 1, 2, \ldots, j\}.$$

Since the system (3.7.9) is not solvable

$$C \cap \left(\bigcap_{i=j+1}^{m} C_i \right) = \emptyset.$$

Without any loss of generality, we can assume that (3.7.9) is solvable if we drop one of the conditions $g_i(x) \geqq 0$. Consequently,

$$C \cap \left(\bigcap_{\substack{i=j+1 \\ i \neq k}}^{m} C_i \right) \neq \emptyset, \quad k = j + 1, \ldots, m.$$

According to the Corollary 1.2.2, Theorem 1.2.8, the convex set C is not contained in $\bigcap_{i=j+1}^{m} C_i$. Thus, there exists an element $x^1 \in C$ with $g_i(x^1) < 0$, $i = j + 1, \ldots, m$ which contradicts the fact that $g(x) = 0$ for any $x \in \mathbb{R}^n$. This contradiction has been constructed on the basis of the assumption $\alpha_0 = 0$. Thus $\alpha_0 \neq 0$ and this proves the theorem.

Application of Farkas-Minkowski theorem will lead us to some statements which are equivalent to statements (i), (ii), (iii) of Theorem 3.7.2.

Let us consider a Lagrangian form

$$\varphi(x, y) = f(x) + \langle y, G(x) \rangle$$

defined on $\mathbb{R}^n \times \mathbb{R}^{m+}$. If x is any element in $\mathbb{R}^n$, then we denote by $I(x)$ a subset of indices $I(x) \subseteqq \{1, \ldots, m\}$ where $i \in I(x)$ if and only if $g_i(x) = 0$.

Now, we formulate the following problems:

(P_1) Problem of a local maximum.

Find a element $x \in M = \{x \in \mathbb{R}^n : G(x) \geqq 0\}$, such that for any $z \in \mathbb{R}^n$

$$f(x + \lambda z) \leqq f(x) \tag{3.7.12}$$

where $0 \leqq \lambda \leqq \lambda_0$ for some value $\lambda_0 > 0$ and $x + \lambda z \in M$.

(P_2) Under the assumption that functions $f, g_1, \ldots, g_m$ are differentiable, find a point $x \in M$, satisfying

$$\langle (\mathrm{d}f)_x, z\rangle \leqq 0 \tag{3.7.13}$$

for any $z \in D$, where

$$D = \{z \in \mathbb{R}^n : \langle (\mathrm{d}g_i)_x, z\rangle \geqq 0 \ \forall\ i \in I(x)\}.$$

(P_3) (Lagrange problem). Assume that functions $f, g_1, \ldots, g_m$ are differentiable. Find elements $(x, y) \in M \times \mathbb{R}^{m+}$, which satisfy

$$\begin{aligned} &\langle y, G(x)\rangle = 0, \\ &(\mathrm{d}f_x) + \sum_{i=1}^{m} \eta_i (\mathrm{d}g_i)_x = 0. \end{aligned} \tag{3.7.14}$$

If the functions $f, g_1, \ldots, g_m$ possess continuous first partial derivatives, then the Fréchet derivatives are equal to their gradients and (3.7.14) changes over to

$$\begin{aligned} &\langle y, G(x)\rangle = 0, \\ &\frac{\partial f(x)}{\partial \xi_j} + \sum_{i=1}^{m} \eta_i \frac{\partial g_i(x)}{\partial \xi_j} = 0,\ j = 1, \ldots, n. \end{aligned}$$

Lemma 3.7.4. *Problems* (P_2) *and* (P_3) *are equivalent.*

Proof: Let $(x, y) \in M \times \mathbb{R}^{m+}$ be a solution of (P_3). Then it follows from

$$\langle y, G(x)\rangle = \sum_{i=1}^{m} \eta_i g_i(x) = 0$$

with $y \in \mathbb{R}^{m+}$ and $g_i(x) \geqq 0$ that for all $i \notin I(x)$ the components η_i must be zero.
Consequently, according to (3.7.14), we get

$$(\mathrm{d}f)_x = -\sum_{i=1}^{m} \eta_i (\mathrm{d}g_i)_x = -\sum_{i \in I(x)} \eta_i (\mathrm{d}g_i)_x.$$

If z is any element contained in D, it follows that

$$\langle (\mathrm{d}f)_x, z\rangle = \left\langle -\sum_{i \in I(x)} \eta_i (\mathrm{d}g_i)_x, z\right\rangle = -\left\langle \sum_{i \in I(x)} \eta_i (\mathrm{d}g_i)_x, z\right\rangle \leqq 0,$$

but this is the inequality (3.7.13).
Now, we assume that x is a solution of (P_2). Then, the system

$$\begin{aligned} &\langle (\mathrm{d}g_i)_x, z\rangle \geqq 0, \quad i \in I(x), \\ &\langle (\mathrm{d}f)_x, z\rangle > 0 \end{aligned}$$

does not possess a solution $z \in \mathbb{R}^n$ while the system

$$\langle (\mathrm{d}g_i)_x, z\rangle \geqq 0, \quad i \in I(x),$$

is soluble; for example $z = 0$ is a solution. According to Theorem 3.7.5, there exist numbers $\lambda_i \geqq 0$, $i \in I(x)$ such that for any $z \in \mathbb{R}^n$, we have

$$\langle (\mathrm{d}f)_x, z\rangle + \sum_{i \in I(x)} \lambda_i \langle (\mathrm{d}g_i)_x, z\rangle = \left\langle (\mathrm{d}f)_x + \sum_{i \in I(x)} \lambda_i (\mathrm{d}g_i)_x, z\right\rangle \leqq 0.$$

This is possible only for

$$(\mathrm{d}f)_x + \sum_{i \in I(x)} \lambda_i (\mathrm{d}g_i)_x = 0 .$$

Now, we choose $y \in \mathbb{R}^{m+}$ with components

$$\eta_i = \begin{cases} \lambda_i, & i \in I(x), \\ 0, & i \notin I(x). \end{cases}$$

Consequently, we get

$$\langle y, G(x) \rangle = \sum_{i \in I(x)} \lambda_i g_i(x) = 0 ,$$

$$(\mathrm{d}f)_x + \sum_{i=1}^{m} \eta_i (\mathrm{d}g_i)_x = 0 ;$$

but, this is (3.7.14).

Thus (x, y) is a solution of (P_3). □

As usual, by a primal problem, we mean the following.

$$\text{Find } W = \max_{x \in M} f(x) \tag{P}$$

with $M = \{x \in \mathbb{R}^n : G(x) \geqq 0\}$.

Lemma 3.7.5. *If $f, g_1, \ldots, g_m$ are concave functions, the problems* (P) *and* (P_1) *are equivalent.*

Proof: Even without the assumption of concavity, obviously every solution of (P) is also a solution of (P_1). Let $x^0 \in M$ be a solution of (P) and $x \in M$ an arbitrary point. Since $g_i(x)$ are concave, we have

$$g_i\big(\lambda x + (1 - \lambda)\, x^0\big) \geqq \lambda g_i(x) + (1 - \lambda)\, g_i(x^0) \geqq 0$$

for any λ with $0 < \lambda < 1$. Consequently, $\lambda x + (1 - \lambda)\, x^0 = x^0 + \lambda(x - x^0)$ is contained in M. According to problem (P_1), we choose λ so small that

$$f\big(x^0 + \lambda(x - x^0)\big) \leqq f(x^0) .$$

f being concave,

$$f(x) + (1 - \lambda)\, f(x^0) \leqq f(x^0) .$$

Thus,

$$f(x) \leqq f(x^0)$$

for any $x \in M$. Consequently, x^0 is a solution of (P). □

Lemma 3.7.6. *If the functions $f, g_1, \ldots, g_m$ are concave and if there exists at least one point $x \in M$ with $g_i(x) \neq 0$ for all nonlinear-affine functions g_i, then the saddle point satisfying inequality* (i) *of Theorem* 3.7.2. *is equivalent to the solution of* (P).

Proof: If $\varphi(x, y)$ possesses a saddle point (x^0, y^0) in $\mathbb{R}^n \times \mathbb{R}^{m+}$, then according to Theorem 3.7.2 saddle-point coordinate x^0 is a solution of (P).

Now we assume that x^0 is a solution of (P). Then, the system

$$g_i(x) \geqq 0, \quad i = 1, 2, \ldots, m,$$
$$f(x) - f(x^0) > 0$$

does not possess a solution.

The assumption of Lemma 3.7.6 permit the application of Farkas-Minkowski theorem (Theorem 3.7.5). Consequently, there exists vector $y^0 \in \mathbb{R}^{m+}$, $y^0 \neq 0$, such that

$$f(x) - f(x^0) + \sum_{i=1}^{m} \eta_i^0 g_i(x) \leqq 0$$

for any $x \in \mathbb{R}^n$. Furthermore, for any $y \in \mathbb{R}^{m+}$,

$$f(x^0) \leqq f(x^0) + \langle y, G(x^0)\rangle = \varphi(x^0, y).$$

From this, it follows that, for any $(x, y) \in \mathbb{R}^n \times \mathbb{R}^{m+}$,

$$\varphi(x, y^0) \leqq f(x^0) \leqq \varphi(x^0, y).$$

Obviously special value $x = x^0$ and $y = y^0$ gives

$$f(x^0) = \varphi(x^0, y^0).$$

Thus (x^0, y^0) is a saddle point of φ. □

Lemma 3.7.7. *If the functions $f, g_1, \ldots, g_m$ are differentiable, then a saddle point (x^0, y^0) of φ is a solution of problem* (P_3).

Proof: Let (x^0, y^0) be a saddle point. From statement (iii) of Theorem 3.7.2 follows $\langle y, G(x^0)\rangle = 0$. Furthermore, the saddle-point inequality implies

$$\varphi(x^0 + \lambda x, y^0) \leqq \varphi(x^0, y^0)$$

for any $x \in \mathbb{R}^n$ and $\lambda > 0$, so that

$$\frac{\varphi(x^0 + \lambda x, y^0) - \varphi(x^0, y^0)}{\lambda} \leqq 0.$$

This is the same as

$$\frac{f(x^0 + \lambda x) - f(x^0)}{\lambda} + \sum_{i=1}^{m} \eta_i^0 \frac{g_i(x^0 + \lambda x) - g_i(x^0)}{\lambda} \leqq 0.$$

Because the functions $f, g_1, \ldots, g_m$ are differentiable and according to the Definition 1.4.2 (of a Fréchet derivative), we get for any $x \in \mathbb{R}^n$ and $\lambda \to 0$,

$$(\mathrm{d}f)_{x^0}(x) + \sum_{i=1}^{m} \eta_i^0 (\mathrm{d}g_i)_{x^0}(x) = \left\langle (\mathrm{d}f)_{x^0} + \sum_{i=1}^{m} \eta_i^0 (\mathrm{d}g_i)_{x^0}, x \right\rangle \leqq 0.$$

This is possible only for

$$(\mathrm{d}f)_{x^0} + \sum_{i=1}^{m} \eta_i^0 (\mathrm{d}g_i)_{x^0} = 0.$$

Because (P_2) and (P_3) are equivalent, it follows from the existence of a saddle point of φ that (P_2) is also solvable. □

Lemma 3.7.8. *If the functions $f, g_1, \ldots, g_m$ are differentiable and if furthermore f is concave, then every solution of* (P_2) *is also a solution of* (P_1).

Proof: Let x^0 be a solution of (P_2). Now we consider the following cases.

Case 1: Consider an arbitrary element $z \in \mathbb{R}^n$ and assume the existence of at least one index $i_0 \in I(x^0)$ such that

$$\langle (\mathrm{d}g_{i_0})_{x^0}, z\rangle < 0 .$$

From this follows the existence of a real number $\lambda_0 > 0$ such that

$$\frac{g_{i_0}(x^0 + \lambda z) - g_{i_0}(x^0)}{\lambda} = \langle (\mathrm{d}g_{i_0})_{x^0}, z\rangle + \frac{a(x^0, \lambda z)}{\lambda} < 0$$

for any $0 < \lambda \leqq \lambda_0$. Thus,

$$g_{i_0}(x^0 + \lambda z) < g_{i_0}(x^0) = 0 .$$

In this case, the element $x^0 + \lambda z$ with $0 \leqq \lambda \leqq \lambda_0$ is contained in M only for $\lambda = 0$.

Case 2: Now assume $z \in D$ where

$$\langle (\mathrm{d}g_i)_{x^0}, z\rangle \geqq 0 \quad \forall\, i \in I(x^0) .$$

Since x^0 is a solution of (P_2), we have

$$\langle (\mathrm{d}f)_{x^0}, z\rangle \leqq 0 .$$

Because f is concave, by Theorem 1.4.10,

$$f(x^0 + \lambda z) - f(x^0) \leqq \lambda\langle (\mathrm{d}f)_{x^0}, z\rangle$$

for any $\lambda > 0$. Consequently,

$$f(x^0 + \lambda z) \leqq f(x^0) + \lambda\langle (\mathrm{d}f)_{x^0}, z\rangle \leqq f(x^0) .$$

Now for both cases, we can find a real number $\lambda_0 > 0$ such that for $0 \leqq \lambda \leqq \lambda_0$ and $x^0 + \lambda z \in M$, it follows that

$$f(x^0 + \lambda z) \leqq f(x^0) .$$

Thus x^0 is a solution of (P_1). □

If the function f is not concave, we can show with a simple example given below that a solution of (P_2) is not a solution of (P_1). Consider $f(\xi) = \xi^2$ defined on $\mathbb{R}$ and $g(\xi) = 0$. Obviously $M = \mathbb{R}$ and for f there does not exist a local maximum. On the other hand, $D = \mathbb{R}$ and $\xi^0 = 0$ is a solution of (P_2). Another example shows that in special cases a solution of (P_1) is not a solution of (P_2).

For $x = (\xi_1, \xi_2) \in \mathbb{R}^2$, consider the functions

$$g_1(x) = \xi_1 ,$$

$$g_2(x) = \xi_2 ,$$

$$g_3(x) = (1 - \xi_1)^3 - \xi_2$$

and find a maximum for $f(x) = \xi_1$ over

$$M = \{x \in \mathbb{R}^2 : g_i(x) \geqq 0\} = \{x \in \mathbb{R}^{2+} : (1 - \xi_1)^3 \geqq \xi_2\} .$$

The sketch in Fig. 8 shows the admissible domain M. Obviously $x^0 = (1, 0)$ is the uniquely defined solution of (P) and (P_1).

Furthermore,

$$g_1(x^0) = 1, \quad g_2(x^0) = 0, \quad g_3(x^0) = 0.$$

Thus, $I(x^0) = \{2, 3\}$ and

$$(\mathrm{d}f)_{x^0} = (1, 0), \ (\mathrm{d}g_1)_{x^0} = (1, 0), \ (\mathrm{d}g_2)_{x^0} = (0, 1), \ (\mathrm{d}g_3)_{x^0} = (0, -1).$$

From this follows

$$D = \{z = (\zeta_1, \zeta_2) \in \mathbb{R}^2 \colon \zeta_2 \geqq 0, -\zeta_2 \geqq 0\} = \{z \in \mathbb{R}^2 \colon z = (\zeta_1, 0)\}.$$

Obviously $x^0 \in D$. But, on the other hand

$$\langle (\mathrm{d}f)_{x^0}, x^0 \rangle = \langle (1, 0), (1, 0) \rangle = 1 > 0$$

showing that x^0 is not a solution of (P_2).

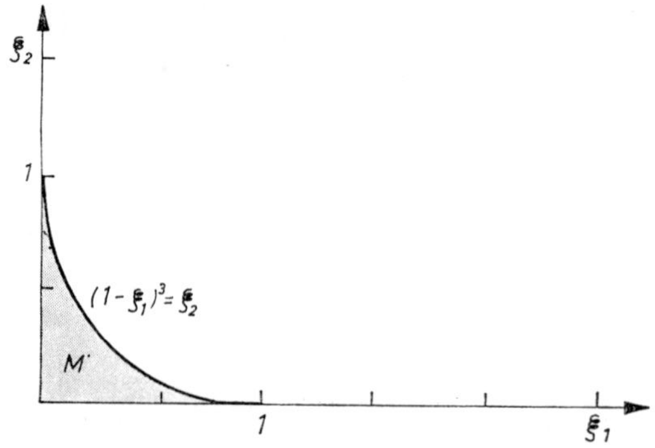

Fig. 8

Now, we can put into relations the Lemmas 3.7.4 to 3.7.8 and Theorem 3.7.2. If we take all the assumptions of Lemmas 3.7.4 to 3.7.8, then all the statements are equivalent. This means the following theorem.

Theorem 3.7.6. *Let the functions $f, g_1, \ldots, g_m$ be differentiable and concave. Furthermore, we assume that there exists at least one point $x \in M$ such that $g_i(x) \neq 0$ for all nonlinear affine functions g_i.*

Then the following problems are equivalent:

(P) Find $W = \max\limits_{M} f(x)$

where $M = \{x \in \mathbb{R}^n \colon G(x) \geqq 0\}$.

(P_1) Find $x^0 \in M$ such that for any $z \in \mathbb{R}^n$,

$$f(x + \lambda z) \leqq f(x)$$

where $0 \leqq \lambda \leqq \lambda_0$ for some real number $\lambda_0 > 0$ and $x + \lambda z \in M$.

(P_2) Find $x^0 \in M$ which satisfies

$$\langle (\mathrm{d}f)_{x^0}, z \rangle \leqq 0, \quad \text{for all } z \in D.$$

(P_3) Find $(x^0, y^0) \in M \times \mathbb{R}^{m+}$ which satisfies

$$\langle y^0, G(x^0) \rangle = 0,$$

$$(\mathrm{d}f)_{x^0} + \sum_{i=1}^{m} \eta_i{}^0 (\mathrm{d}g_i)_{x^0} = 0.$$

(I) Find $(x^0, y^0) \in \mathbb{R}^n \times \mathbb{R}^{m+}$ such that

$$f(x) + \langle y^0, G(x)\rangle \leqq f(x)^0 + \langle y^0, G(x^0)\rangle \leqq f(x^0) + \langle y, G(x^0)\rangle$$

for any $(x, y) \in \mathbb{R}^n \times \mathbb{R}^{m+}$.

(II) Find $(x^0, y^0) \in M \times N$ where

$$f(x^0) = \max_M f(x) = \min_N v(y) = v(y^0).$$

(III) Find $(x^0, y^0) \in M \times N$ where

$$f(x^0) = v(y^0)$$
$$\langle y^0, G(x^0)\rangle = 0.$$

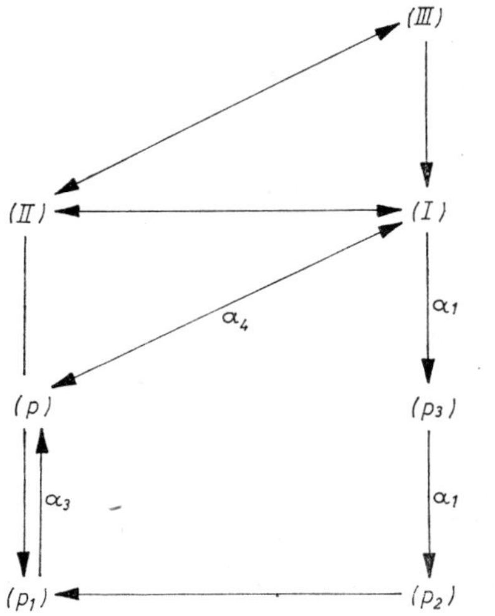

Fig. 9

The following schematic diagram shows the relations between Theorem 3.7.2 and Lemmas 3.7.4 to 3.7.8. To draw the diagram we introduce the following notations.

(α_1): functions $f, g_1, \ldots, g_m$ are differentiable,
(α_2): functions $f, g_1, \ldots, g_m$ are differentiable and f is concave,
(α_3): functions $f, g_1, \ldots, g_m$ are concave,
(α_4): functions $f, g_1, \ldots, g_m$ are concave and there exists at least one $x \in M$ where $g_i(x) \neq 0$ for all nonlinear-affine function g_i.

If we can go over from one problem to another without demanding any more properties of the functions f and g_i, we just draw a link and denote the direction with an arrow. If we need one of the conditions $\alpha_1, \alpha_2, \alpha_3, \alpha_4$, to go over from one problem to another, we label the link with α_i. Thus we have the schematic diagram in Fig. 9.

Literature

[1] BELLMAN, R.: Dynamic Programming, Princeton University Press, Princeton 1957.

[2] BERGE, C.: Sur une propriete combinatoire des ensembles convexes, C. R. Academie des Sci., Paris, 248, (1959) 2698—2699.

[3] BERGE, C.: Théorie des graphes et ses applications, Dunod, Paris 1958.

[4] BERGE, C., and A. GHOUILA-HOURI: Programming, Games, and Transportation Networks German transl.: (Programme, Spiele, Transportnetze, BSG B. G. Teubner Verlagsgesellschaft, Leipzig 1969).

[5] DIETER, U.: Dualität bei konvexen Optimierungs- (Programmierungs-) aufgaben, Unternehmensforschung **9** (1965), 91—111.

[6] DIETER, U.: Optimierungsaufgaben in topologischen Vektorräumen, Zeitschr. f. Wahrscheinlichkeitsth. **5** (1966), 89—117.

[7] DRAGAN, I.: An optimality condition for minimal-cost multicommodity flows, in: Math. Progr. Activity Analysis, Amsterdam, (1975), 135—139.

[8] DUFFIN, R. J.: Linear inequalities and related systems, Ann. of Math. **38** (1956), 171—183.

[9] EDWARDS, R. E.: Functional Analysis — Theory and Application, Holt, Rinehart and Winston, New York 1965.

[10] FAN, K.: Fixed point and minimal theorems in locally convex topological linear spaces, Proc. Nat. Acad. Sc. USA **38**, No. 2 (1952), 137—152.

[11] FARKAS, J.: Über die Theorie der einfachen Ungleichungen, J. reine und angew. Math. **124** (1902), 1—24.

[12] FENCHEL, W.: On conjugate convex functions, Can. J. Math. **1** (1949), 73—77.

[13] GLICKSBERG, J.: A further generalization of the Kakutani fixed point theorem with applications to Nash equilibrium points, Proc. Amer. Math. Soc. **3** (1952), 170—174.

[14] GOLSTEIN, E. G.: Dualitätstheorie in der nichtlinearen Optimierung und ihre Anwendung, Akademie-Verlag, Berlin 1975.

[15] GOLSTEIN, E. G. Двойственные задачи Быпуклого и Дробно-Выпуклого программирования, В сб. ,,Исследования по математическому программированнию". Nauka, Moscow 1968, 10—108.

[16] GÖPFERT, A.: Mathematische Optimierung in allgemeinen Vektorräumen, BSB B. G. Teubner Verlagsgesellschaft, Leipzig 1973.

[17] HADLEY, G.: Nonlinear and Dynamic Programming, Addison-Wesley, Reading, Massachusetts 1965.

[18] IOFFE, A. D. and V. M. TIKHOMIROV (Иоффе, А. Д. и Тихомиров, В. М.) Двойственность выпуклых функций и экстреммльныд задачи. УМН **23**, 6 (144) (1968), 1-65.

[19] KAKUTANI, S.: A generalization of Browers fixed point theorem, Duke Math. J. 8 (1941), 457—459.

[20] KARLIN, S.: Mathematical Methods and Theory in Games, Programming and Economics, Vol. I, Addison-Wesley, Reading, Massachusetts 1960.

[21] KARLIN, S.: Operator treatment of minimax principle, Ann. Math. **24** (1950), 133—154.

[22] KREIN, M. G.: On regular convex sets in the space conjugate to a Banach space, Ann. Math. (2) **41** (1940), 556—583

[23] KREKO, B.: Lehrbuch der linearen Optimierung, Dt. Verl. d. Wiss., Berlin 1964.

[24] KRETSCHMER, K. S.: Programs in paired spaces, Can. J. Math. **13** (1961), 221—238.

[25] KUHN, H. W., and A. W. TUCKER: Nonlinear Programming, Proceedings of the Second Berkley Symposium on Math. Stat. and Prob., Berkley (1951), 481—492.

[26] Loomis, L. H.: On a theorem of von Neumann, Proc. Nat. Acad. Sc. USA **32** (1946), 213–215.
[27] Moreau, J. J.: Sur la fonction poliare d' une fonction semicontinue superienrement, Compt. Rend. Acad. Sci. **258** (1964), 1128–1130.
[28] Moreau, J. J.: Fonctions convexes en dualite, Faculte des Science de Montepellier Seminaire de Math. (Multigraph) (1962).
[29] Moreau, J. J.: Fonctions convexes duals et points proximaus dans une espace Hilbertian, Compt. Rend. Acad. Sci. **255** (1962), 2897–2899.
[30] Neumann, J. v., and O. Morgenstern,: Theory of games and economic behavior, Princeton 1944 (German transl.: Spieltheorie und wirtschaftliches Verhalten, Physica-Verlag, Würzburg 1961).
[31] Sion, M.: Sur une generalisation du theoreme minimax, C. R. Academie des Sci. **244** (1957), 2120–2123.
[32] Vogel, W.: Lineares Optimieren, Akad. Verlagsges., Geest & Portig, Leipzig 1970.
[33] Vogel, W.: Duale Optimierungsaufgaben und Sattelpunktssätze, Unternehmensforschung **4** (1969), 1–28.
[34] Walk, M., and C. Swe: Theory of Duality, University text Arts and Science University, Rangoon 1975.
[35] Walk, M.: Optimale Programme für Multiströme auf endlichen gerichteten Graphen, Wiss. Z. Friedrich-Schiller-Univ. Jena, Math.-Nat. R., **29** (1980), 319–325.
[36] Walk, M.: Dualität eines verallgemeinerten Transport- und Potentialproblems, Math. Operationsf. Statist., Ser. Optimization, **11**, No. 3 (1980), 465–472.
[37] Walk, M. (Вальк, М.): Двойственност транспортной и потенциальной задач в банаховых пространствах, труды тбилисского университета, A 9, **157** (1975) 13–19.
[38] Walk, M. (Вальк, М.): Обобщенные потоки и напряжения на графах, Вестник БГУ, серия I, физика, математика, механика, **2** (1985), 51–55.

Notation index

$\bar{C}$	closure of C
C^*	conjugate set to C
Conv (A)	convex hull
$\overline{\text{Conv}\ (A)}$	closure of the convex hull
$(\mathrm{d}f)_{x^0}$	Fréchet differential, total derivative
$\mathrm{D}_i f(x^0)$	partial derivative
E	Banach space
E^*	dual Banach space
E^{**}	bidual space
f^*	conjugate function of the function f
grad	gradient
H	hyperplane
H^+	positive half space defined by hyperplane H
H^-	negative half space defined by hyperplane H
L_2	Hilbert space of quadratic integrable function
$L_2{}^+$	nonnegative of L_2
$L_2{}^n$	$= L_2 \times L_2 \times L_2 \ldots \times L_2$
$\mathbb{R}$	set of real numers
$\mathbb{R}^n$	n-dimensional Euclidean space
$\mathbb{R}^{n+}$	convex cone $\{x \in \mathbb{R}^n : x > 0\}$
sign ξ	$= \begin{cases} 1 & \text{for} \quad \xi > 0 \\ 0 & \text{for} \quad \xi < 0 \end{cases}$
P, Q	convex cone
P*, Q*	dual cone
T^*	transposed matrix of the matrix T
$[T, P, Q, a^*, b]$	primary program
$[T, P, Q, a^*, b]^{\circledast}$ $= [-T^*, Q^*, P^*, -b, a^*]$	dual program
$\langle a, x \rangle$	scalar product
$\lVert x \rVert$	Euclidean norm
$\square$	end of the proof
$\delta(C_1, C_2)$	distance between C_1 and C_2
$\emptyset$	empty set

Subject index